Morocco

Globalizing Regions

Globalizing Regions offers concise accounts of how the nations and regions of the world are experiencing the effects of globalization. Richly descriptive yet theoretically informed, each volume shows how individual places are navigating the tension between age-old traditions and the new forces generated by globalization.

Australia: Nation, Belonging, and Globalization — Volume One
Anthony Moran

Global Hong Kong — Volume Two
Gary McDonogh and Cindy Wong

On Argentina and the Southern Cone: Neoliberalism and National Imaginations — Volume Three
Alejandro Grimson and Gabriel Kessler

The Koreas — Volume Four
Charles Armstrong

China and Globalization: The Social, Economic, and Political Transformation of Chinese Society —
Volume Five
Doug Guthrie

Morocco: Globalization and Its Consequences — Volume Six
Shana Cohen and Larabi Jaïdi

Forthcoming:

Global Iberia — Volume Seven
Gary McDonogh

Ireland — Volume Eight
Tom Inglis

The Globalization of Israel: McWorld in Tel Aviv, Jihad in Jerusalem — Volume Nine
Uri Ram

Global Indonesia — Volume Ten
Jean Gelman Taylor

Global Iran — Volume Eleven
Camron Michael Amin

Morocco

Globalization and
Its Consequences

SHANA COHEN

AND

LARABI JAIDI

Routledge
Taylor & Francis Group
New York   London

Routledge is an imprint of the
Taylor & Francis Group, an informa business

Routledge
Taylor & Francis Group
270 Madison Avenue
New York, NY 10016

Routledge
Taylor & Francis Group
2 Park Square
Milton Park, Abingdon
Oxon OX14 4RN

© 2006 by Taylor and Francis Group, LLC
Routledge is an imprint of Taylor & Francis Group, an Informa business

Printed in the United States of America on acid-free paper
10 9 8 7 6 5 4 3 2 1

International Standard Book Number-10: 0-415-94511-9 (Softcover) 0-415-94510-0 (Hardcover)
International Standard Book Number-13: 978-0-415-94511-0 (Softcover) 978-0-415-94510-3 (Hardcover)
Library of Congress Card Number 2005035976

Library of Congress Cataloging-in-Publication Data

Cohen, Shana, 1966-
    Morocco : globalization and its consequences / Shana Cohen and Larabi Jaïdi.
          p. cm. -- (Globalizing regions)
    Includes bibliographical references and index.
    ISBN 0-415-94510-0 (hb) -- ISBN 0-415-94511-9 (pb)
    1. Morocco--Politics and government--1999- . 2. Morocco--Social policy.
3. Morocco--Economic policy. 4. Islam and state--Morocco. 5. Globalization.
I. Jaïdi, Larbi. II. Title. III. Series: Globalizing regions series.

DT326.3.C64 2006
338.964'053--dc22   2005035976

Visit the Taylor & Francis Web site at
http://www.taylorandfrancis.com

and the Routledge Web site at
http://www.routledge-ny.com

Contents

Preface ix

One
Debating and Implementing "Development" in Morocco 1

Two
Who Cares about Political Reform? 47

Three
The Power of Political Movements 79

Four
Assessing the Social Impact of Development 113

Conclusion 151

Endnotes 157

Bibliography 169

Index 175

—a mes amis et mes proches

L.J.

—for the two Rs

S.C.

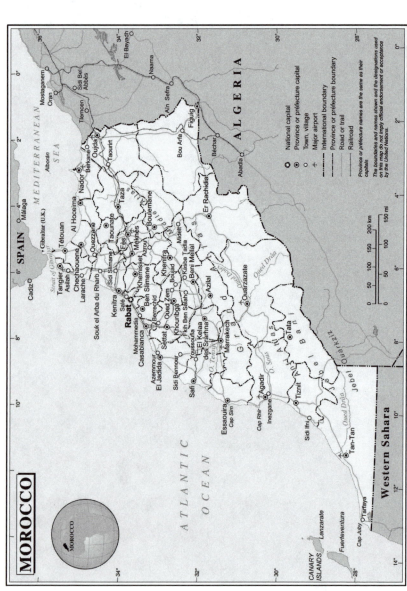

MOROCCO

Preface

This book evolved out of research and years of experience working with development projects and agencies. In Larbi Jaïdi's case, these years of experience—far more than Shana Cohen's—also include leadership within the Socialist party, the Union Socialiste des Forces Populaires (USFP) party, and advisory positions to government officials. For both of us, academic work has remained paramount. One of our central concerns, though, has consistently been practical. We have both explored how market reform has affected income distribution, the division of labor, and general patterns of inequality. We have asked how we determine the causes of inequality. How do we define and measure this inequality, and what are its consequences? What can we do about it through economic and social policy?

These questions are particularly relevant in Morocco, which, as political figures and officials in development agencies often note, possesses relatively poor social indicators relative to national income. Morocco is ranked below other countries in the Arab world in major social indicators, including female literacy and primary school completion rates and, importantly, below other countries globally in the same income category (i.e., lower middle income). For instance, female illiteracy in

Morocco in 2000 was 63.9% of the total population, whereas the average for lower middle income countries was 19.4%.[1]

Taking our ethical and intellectual interests into consideration, when we were commissioned to write a book about Morocco and globalization, we elected to emphasize the process of "development." For a low-middle-income country like Morocco, globalization implies economic and political reform as well as cultural consumption. The book likewise began as a reflection on how Morocco had changed socially, politically, and culturally with the implementation of market reform policies. As we researched and wrote the book, we organized our analysis differently. Globalization as development in Morocco has created tensions and contradictions that plague a state managing both economic liberalization as well as poverty and unemployment. The state has also engaged in political liberalization, allowing for greater freedom of expression and a wider range of opposition activity and mobilization. Accordingly, as the king and successive governments have implemented reforms, they have faced growing criticism and pressure from actors outside of government, ranging from women's organizations to Islamist parties to an expanding press.

In our previous work, we have explored these tensions and contradictions as symptomatic of larger social and political changes. Cohen has written about the transformation of the middle class in Morocco from a modern middle class attached to modernization policies to a global middle class created through the processes of global market integration (2003, 2004). Larbi Jaïdi has written in various articles for the Moroccan press about the need for policy to address persistent inequality, and the disaffection of youth and economically and socially marginalized populations, in order to promote

Morocco's progress as a country and to prevent social revolt. Our work reclaims the urgency of policy intervention for economic growth and income distribution at a time when Islamic radicalism and widespread disaffection permeate the region. Importantly, this intervention must confront structural social problems and a serious political impasse, and not settle for temporary solutions.

As a country, Morocco opened its borders to capital and advice for a relatively short period—primarily since the nineteenth century. Morocco was never subjugated to Ottoman rule like the rest of the region and held off European influence long after Algeria fell increasingly under French control after 1830. Instead, Morocco's tribal structure, religious tradition of *maraboutism* (Islamic mysticism centered on saints and brotherhoods), and the sultanate contributed to the independence and integrity of the country and its capacity to contest foreign enemies, from the Turks to the Spaniards. In addition, Morocco benefited from continuity in the sultanate, if not in the particular dynasty at least in the institution. And Morocco in the precolonial period still claimed the heritage of the glorious Andalusian period: the renaissance of Islam in the Middle Ages under Berber-Moroccan dynasties such as the Almohades and the Merinides.

The current dynasty of the Alaouites came to power in 1666, making them one of the longest-ruling families in world history. They clung to power from the 1600s through the mid-1700s by occasionally imposing taxation, building armies, and countering the localized power of brotherhoods with Sharifism—a sect emphasizing lineage from the Prophet—which the Alaouites also claimed. In the mid-1700s, Sultan Mohammed III eventually transcended the weaknesses of the state (called *makhzen*), which had become bloated and parasitic, by decentralizing the administration and encouraging commerce to supplement the state's income. He also deliberately weakened the power of the army, which had repeatedly tried to overthrow his grandfather, Sultan 'Abd Allah.

Successive sultans managed the foreign relations introduced and accelerated by Mohammed III in the second half of the 1700s through diplomacy and trade, mainly conducted by the Moroccan Jewish population. These sultans after Mohammed III also tried to maintain the continuation of piracy, which provided the state with income. Yet, piracy, as well as increasing state expenditures and trade, was to provide European countries with reasons for intervention into Moroccan affairs. The French succeeded in repressing piracy in 1844, and the English signed a trade agreement with Morocco in 1956 that abolished the sultan's trade monopolies, with the exception of tobacco and opium.

As European nations expanded their economic investment in Morocco, they demanded greater extraterritorial privileges, or immunity from Moroccan legislation. This extraterritorial privilege extended to the Moroccan business partners of Europeans (often Jews), making their own consulates in essence autonomous states within a state and possessing their own independent authority to regulate business. According to Abdallah Laroui, perhaps Morocco's most prominent historian of the past century, "Every advantage accorded the foreign powers provided them a permanent means of intervention. As in Algeria and Tunisia, the slogan of freedom (of trade, of property, of the individual) served to undermine the state from within and prepare the way for the introduction of the capitalistic system" (1977, p. 320).

More than capitalism, encroaching European power over business and law in Morocco established the foundation for colonialism. The British agreed to protect the power of the sultanate under 'Abd al-Rahman if he implemented reforms favorable to the wealthy and onerous to the poor, actions that weakened his authority and prestige among his population. At the same time, the sultan's circle of friends elected to become allies of Europeans rather than champions of national independence. From 'Abd al-Rahman's tenure as sultan (1822 to 1859) until the establishment of the French protectorate in 1912, the sultans became trapped by their disintegrating authority over the population and their own institution. As Laroui comments, "By 1880 the Moroccan state had ceased to be anything more than a fiction" (1977, p. 322).

French colonialism accelerated commerce and industrialization in Morocco, creating an urban working class and a nascent middle class. The French, under the direction of General Lyautey, also pursued a "divide and conquer" strategy. This strategy focused particularly on distinguishing Berbers from Arabs in religion and cultural heritage. The Protectorate removed Berbers from the authority of the Makhzen and encouraged the establishment of separate institutions such as the Berber College. This policy was articulated in the Berber Dahir (a legal instrument utilized by the sultan) of 1930.

While the French implemented policies designed to ensure their control, they also fostered the bases for resistance. French unions created unions in Morocco, the growth of education produced a new professional and intellectual class, and French trade and investment catalyzed Moroccan business. This latter group, however much they benefited from French colonization, resented the blocked mobility that foreign rule implied—that a Moroccan could never attain the status or power of a Frenchman. In addition, in the efforts of the French to develop modern institutions, they alienated the *alims*, the Muslim clergy, and traditional local leaders. The clergy and local leaders turned to a conservative Islam, one "purified" of distortion, to liberate the country in general and the sultan in particular from foreign influence. This turn to Islam became *salafisme*, which influenced heroic nationalist leaders such as Abdelkrim al-Khattabi, who led the rebellion in the mountainous Rif region (1919 to 1926) and Allal al-Fassi, who became one of the leaders of the nationalist Istiqlal party. Resistance against colonialism took the form of restoration of Muslim, Arab identity, and law.

Al-Fassi referred to the role of identity in his experience as a nationalist leader and thinker:

> None of us, whatever the economic-social system that we preferred, was able to imagine a Morocco that was neither Arab or Muslim. I belong to one of these families that lived seven centuries in Andalusia before the vanquishing Christians refused them the right of practicing their religion freely, obliging them to emigrate. Since then, we have never ceased to feel nostalgia for this land of a glorious past and to hope that God will deliver it from its misfortunes, but we do not think of her (medieval Spain) as our country. This, for us, is Morocco and only Morocco, as if we had never emigrated elsewhere, because it is here that we found a "repository for the body and for the soul" in the sense of verse 59 [of the Koran].[2]

Into this simmering nationalist fervor stepped the sultan, Mohammed V, the grandfather of the current king. Defying direct rule installed after the departure of Lyautey in 1925, the sultan took steps to assert his independent authority and Morocco's territorial integrity. He refused the order under Vichy France to deport Moroccan Jews. He also declared independence in 1944 and his sovereignty over Tangiers in 1947. Negotiations with the French authority ended in stalemate, and in 1952, Mohammed V was exiled with his family to Madagascar. He returned in 1955 to achieve real independence in 1956.

Mohammed V did not live long enough (he died in 1961) to face the difficulties of modernizing an economy and participating in the world system of nation-states. His son, Hassan II, assumed this responsibility, often using terribly repressive measures to maintain his power. When he died in 1999, he left to his son, the current King Mohammed VI, the obligations and pressures of our own era of globalization.

This book picks up here, in the current era of neoliberalism and shaky political reform. The book combines sociological and political analysis in interpreting the effects of contemporary development policy upon the social fabric and political life of Morocco.

In exploring the tensions and contradictions that have emerged through processes of political and economic liberalization, we extend our analysis outward, from changes within political and economic institutions to the social and political trends occurring as the consequence of these institutional changes. In the first chapter, we provide background on the institutions and organizations that work in development and participate in the major debates. We then outline how

Morocco, a stellar student of development policy trends, has implemented policy since independence. We show at the end of the chapter that despite adopting the recommendations of the World Bank and IMF and the presence of development agencies from the Peace Corps to UNDP, Morocco still has high rates of poverty, unemployment, and illiteracy.

In the second chapter, we focus on the impact of the process of political liberalization. Although characterized by the contradictions mentioned earlier, such as expansion of the press alongside incidents of censorship and imprisonment, political action outside the authority of the king and the state in general has indeed become more possible. We outline three responses to political liberalization. The first is more dynamism within political institutions and civil society, though both domains remain dominated by elites. The second is the rise of cultural–political movements that link transnational political networks and organizations with local activism. The most prominent of these movements are the Islamists, the women's movement, and the Amazigh movement. The third response is widespread disaffection caused by the weakening connection between nonelites and the state.

In the third chapter, we examine the cultural-political movements in more depth. We show how they have surpassed in power and influence the radical leftist movements of the 1960s and 1970s and even the conservative *salafiyya* nationalist Islam advocated by the party of independence, Istiqlal. The Islamists do not represent a unified bloc and, indeed, their divisions reflect the distinctions within political life as a whole. The Parti de la Justice et du Développement (PJD) has elected to participate formally in elections and parliament. In contrast, the largest Islamic party, Al-adl Wal-ihsan (justice and charity), does not participate in formal politics. It takes advantage of its exclusionary status to promote particular religious beliefs, often engaged with mysticism, and to promote the idea of an Islamist state. At the extreme, the radical groups operate as a challenge to the authority of the state, using resources from the Middle East and transnational networks, including al-Qaeda, to further their conservative, highly confrontational interpretation of Islam.

The chapter relates how the Islamist parties and organizations, the women's movement, and the Amazigh (Berber) movement have clashed on issues from reform of the family code, the Moudawana, to the legitimacy of a Berber political party. Despite their substantive differences, though, they share similar tactics of mobilization and occasionally structurally similar arguments for change. The chapter discusses these tactics, from citing international agreements to delivering social services, and indicates their relative success. The last section of the chapter distinguishes the influence and power of these movements, particularly in their ability to challenge the state and the monarch, from radical groups. These groups operate on a more transnational level than the cultural–political movements but fight a more modern fight, without ambiguity or manipulation.

The fourth chapter goes beyond the political analysis of the second and third chapters to investigate social change during the contemporary period of market reform. Although participation in political discourse has grown, and democratic, nongovernmental, and Islamist groups have demonstrated impressive capacity at mobilization, social division has arguably expanded over the past several decades. This division is not based solely on income distribution but also on the commonality, or lack thereof, among different social groups. Factors ranging from different access to health care and education to government policy imperatives have separated out populations. This chapter explores the social impact of market reform policies as a way of evaluating their success. Returning to our underlying interest in inequality, if social groups, from rural low-income communities to wealthy businesspeople, experience "development" as a fragmented, differential process, what are the implications for overall progress as a country?

One

MOROCCO: THEN AND NOW

In an interview with the French newspaper *Le Figaro* in 2001, two years after his ascent to the throne, King Mohammed VI explained his perception of his role as monarch. Young for a head of state at 38 and evidently different from his father, Hassan II, in personality and ambition, Mohammed VI faced expectations that he would modernize the monarchy. He would allow for a transition to democracy and the constitutionalization of royalty. Instead, in the interview he insisted that "Moroccans want a strong monarchy, one that is democratic and executive." Differentiating the monarchy of Morocco from that of Spain and distinguishing himself from King Juan Carlos, he stated, "In Morocco, the King is not content to reign. I reign and I work with my government in a clear constitutional framework that defines the responsibility of each. There is no ambiguity and no complexity in what I am saying to you. For the thirteen centuries that the Moroccan monarchy has endured, we have evolved within this framework and Moroccans want it as well."[1]

Four years later, in 2005, the King conveyed a seemingly different conception of the role of the monarch vis-à-vis his people. In his speech of August 20, 2005, rather than make the comparison with other countries, he distinguished between his own rule and that of his father and grandfather. He called his grandfather, Mohammed V, the "supreme guide of the revolution" who led resistance against the French and was a key figure in the subsequent modernization of the country. His father, Hassan II, created a foundation for a democratic and constitutional monarchy open to political and economic liberalization. His own role in Moroccan history, Mohammed VI elaborated, was to "breathe a new dynamic and to set down the bases for a qualitative transition." He added, "In addition, we have worked without relent on reaffirming a state of law and of institutions, promoting the values of a responsible citizenship, the modernization of the economy, and the concretization of the spirit of solidarity, all for assuring the practice of a democracy with all of its social and human benefit."[2]

What has happened to cause such a shift, particularly from the "executive" monarch to that of "responsible citizenship"? Does the shift represent serious intent to instigate a "qualitative transition" in the position of the monarch, or do the two perceptions of the role of the monarch reflect the contradictions and tensions that characterize contemporary politics in Morocco? The first conception, given in the interview with *Le Figaro*, evokes the monarch of postcolonial Morocco, or an authoritarian leader controlling institutional reform and political participation. The second conception suggests the emergence of a state that responds to the needs of its citizens and, importantly, a monarch that rules citizens rather than governs subjects.

This book attempts to explain the transformation in discourse and its meaning in relation to contemporary development policy, which pushes for economic and political liberalization. We look at how the ideological debates, fund transfers, and policy agendas that make up the contemporary global industry of "international development" have transformed political and social life in a particular place—Morocco. Conversely, we ask how changes in material conditions, political tensions, popular mobilization, public institutions, and so on, become the context for this increasingly dominant version of "development."

The context is important for understanding both the immediate impact of specific policies and programs and the sustainability of the trajectory of political and market liberalization. Perhaps more importantly, focusing on the metamorphosis of a place during the process of global market integration reveals the unpredictability inherent in such a process. Analyzing political and social life during a period of economic and political liberalization highlights the contradictions and tensions of "development" and negates the possibility of a singular path to a determined goal. The number of actors with competing resources and visions makes political conflict and resolution a process of management rather than enforcement. Social problems—such as poverty and the status of women in Morocco's case—under international scrutiny make policy a global enterprise rather than simply a national one.

The most apparent contradiction in Morocco is the dual strength of radicalism and democratization. The process of democratization dates to Hassan II, who, faced with blockage in a system weighted down by patrimony and a legacy of authoritarianism, began to loosen gradually his tight control over political participation and diversity. Mohammed VI has continued along the same path, promoting the most profound process of political liberalization in the region. The state in Morocco has held increasingly transparent elections, permitted greater freedom of expression, and encouraged, particularly through the cultivation of a nonprofit sector (*secteur associatif* in French) and nongovernmental organizations (NGOs), a greater plurality of political actors. Morocco has also made the transition from one monarch to another and from a more closed political system to a more open one without significant destabilization. Malika Zeghal comments that the father and son have made "the Moroccan case ... exceptional in a region where authoritarianism remains a primary characteristic of political regimes."[3]

At the same time, however, Moroccan terrorist cells have been implicated in multiple attacks, Islamist parties have become more popular and more powerful, opposition groups ranging from Amazigh activists to militant leftists have championed particular causes, and political alienation and disaffection have become common, especially among the younger generations. Despite reforms, the political system has not necessarily become more rational or more effective owing to, among other factors, the rise of Islamism (political Islam) and budgetary constraints. Indeed, Zeghal qualifies her analysis of Moroccan exceptionalism by adding, "[Morocco] finishes by rejoining, through the incontrovertible presence of Islamism, the rest of the Arab and Muslim world."[4]

The young King himself has acted in contradictory ways. For instance, at the beginning of his reign, the charismatic Islamist leader Cheikh Yassine sent Mohammed VI a letter, as he had done with Hassan II. As in his previous letter to Hassan II, Yassine instructed Mohammed VI on how to reign according to Islamic principles. In deciding how to respond, Mohammed VI sought the advice of his political advisors, who asked him to ignore the letter (Hassan II, on the other hand, had Yassine confined to an asylum) (see Munson 1993).

Six years later, in 2005, or at the same time Mohammed VI was redefining his position as monarch, he reacted strongly when Yassine's daughter questioned the relevance of the monarchy for Morocco. Mohammed VI quickly had her arrested. Reverting to the back-office tactics of the past, he then negotiated with Yassine's party, Al-adl wal-ihsan, to delay Nadia Yassine's hearing. In other words, he moved to constrain public discussion of the issue.

This complexity of social and political change that occurs during the process of market reform is often missed by the Western media. For example, an article in the *Washington Post* after the terrorist attacks in Casablanca in May 16, 2003, states that, "Moroccan government officials tout the arrests and the absence of additional attacks as evidence that they have neutralized the threat of terrorism. But officials in nearby European countries have expressed fears that Morocco, a country with a tradition of Islamic moderation, is becoming more radicalized."[5]

Radicalization, commentators explain, originates in poverty and political repression. The *Washington Post* article begins by recounting that "King Mohammed VI lifted the hopes of his most impoverished subjects last year when he toured Casablanca's sprawling slums, home to a dozen suicide bombers who had blasted targets across the city. The monarch said he was appalled at the conditions and vowed to raze the shantytowns, promising new housing for an estimated 150,000 people." Yet, the article relates, "Almost 18 months later, the tin-roofed shacks and squatters' colonies are still here. While a few families have been relocated, the most visible change is a freshly built police station that keeps a closer eye on the slums...."[6]

Can we, however, reduce the desire to join an Islamist group and the will to use violence against others and one-self to abstract notions of poverty, political repression, alien-ation, and social inequality? A more profound understanding of what has generated political alienation and encouraged membership in Islamist movements would begin by analyzing how factors ranging from the political power of international and local actors to relations between different social groups have altered with global market integration. Factors such as these shake the foundations of what we have come to think of as a modern society created through policies of modernization and nationalism. Both Islamism and increased consumerism represent responses to the shift initiated in the past several decades under the aegis of economic and social progress.

In analyzing the impact of global market integration as a strategy of development, we can ask basic questions such as whether or not the push for market reform, and thus open competition, has diminished the power of certain economic or political groups. We can also ask questions more difficult to answer: (1) Has the diversification of voices in the inter-national arena challenged identification with the state and altered the hierarchy of authority and control? (2) Who do the voices include: the World Bank, Islamist organizations, West-ern social movements, Western and Arab media, and migrant workers returning from Europe? (3) How do local leaders and individual men and women integrate the voices into a plan of political and social action, or do they?

More explicitly, whereas government actors in Morocco interact with officials from international financial institu-tions (IFIs) and leaders of Western government aid agencies or branches within the United Nations (UN), leaders of political movements and community organizations have their own contacts within global networks and organizations. Every political actor, including the various Islamist parties and groups in Morocco, can utilize global resources and the universalist discourse of rights and progress to promote their own particular vision of advancement in Morocco. The consequence is new terrains of political conflict and mobilization. The state, specifically the monarch, can no longer impose a vision of authority to the detriment of all others, as King Hassan II did for three decades after its independence in 1956.

For example, when Nadia Yassine faced charges of anti-monarchism for questioning the relevance of a monarchy for Morocco, she defended herself by claiming the right to freedom of expression as an individual. This stance provoked statements from the American government and the most liberal faction of Moroccan media supporting the same freedom of expression. She thus forced the hand of the monarch without fearing brutal repression or even the dissolution of the party she represents, actions King Hassan II would have taken twenty years earlier.

Conversely, when the state now takes the initiative to demonstrate authority over the moral direction and religious faith of the country, it often confronts criticism from the local media, foreign organizations and media, and the Moroccan public at large. For example, in 2003, fourteen men between the ages of 22 and 35 received prison sentences of up to a year for playing "heavy-metal" music and wearing clothing bearing symbols resonant of Satanism.[7] Accused of undermining Muslim faith and "playing in public, a music contrary to good manners,"[8] the young men had to defend in court their choice of the café where they met, the black T-shirts they wore, the few words of English they used in the songs they wrote, and the skull-shaped ashtrays and plastic snakes they kept in their homes. When the judge asked one of the men why he preferred a skull-shaped ashtray to any other style or shape, the accused man replied, "It helps me with my studies." The judge then stated, "A normal man does not have something like this in his home." The judge then displayed designs drawn by the same man and asked him, "Why do you waste your time drawing comic strips?"

One of the lawyers for the defense argued at the end that the judge determined guilt without evidence or even presumption of innocence:

> In my opinion, you have not posed the only question that was necessary to pose. As these young men are accused of having undermined the faith of Islam, can we know who was undermined? Do you have names, concrete examples? ... These are our children that we judge ... Monsieur, the prosecutor, you went too far. You speak of crime, and you forget the presumption of innocence. These youth talk about the devil, and so what! He is referred to numerous times in the Qur'an. All your dossier has is what these men have told you. There is no investigation. The dossier is empty.[9]

His opinion echoed that of the liberal media in Morocco, human rights organizations, the families of the accused, music magazines, and organizations worldwide and, most importantly, young men and women. *Tel Quel*, the self-declared defender in the media of youth culture in Morocco (although its staff is criticized for belonging to the same elite it attacks), called the suppression of music a mistake for the state, because "music is a form of resisting extremism and fanaticism." These youth, according to the magazine, demonstrate that despite "the misery in which they live, the area where they were born, are not all human bombs. Give them a chance and their talent will explode."[10] This image may not reflect accurately the socioeconomic position of all of the young men, as some come from middle-class families. However, the point itself holds true; instead of cultivating arenas for self-identification and creative expression that compete with radicalism, the state elected to pursue a self-destructive, regressive legal process.

In the case of the musicians, the state blundered, at least according to certain sections of the Moroccan population and their global allies. Rather than proving its authority over religious practice and the law, the state placed its decisions and its fundamental legitimacy more in doubt. The state's actions also revealed the general difficulty in negotiating political and social change between Islamists and liberals and in negotiating the recommendations of the World Bank while not offering at least short-term solutions for disaffected, unemployed youth. As all these actors adjust and react to one another, they complicate the process of development. It becomes a matter not only of setting policy according to the newest idea, but also of negotiating a fault line of cultural, economic, and political tensions. No single strategy course, whether it be institutional reform or more open markets, can clarify the future or make the present more simple.[11]

This chapter first discusses the shift in dominant development ideology from modernization and import substitution to neoliberalism and institutional and legal reform. The chapter then describes briefly the current context of the global development industry and how at least some organizations have participated in the "development" of Morocco. The chapter ends with a summary of development policies and their impact since independence from the French in 1956.

FROM MODERNIZATION TO UNIVERSALISM

After achieving independence from colonialism, newly independent states largely adopted strategies of import-substitution and modernization. They nationalized resources, industry, and financial institutions. They wanted to promote local over foreign business, build a local market for local goods and, importantly, strengthen and expand the public sector. States varied in their adoption of import-substitution, from the autarkic policies of India to the more open economies of South America. States also invested in education and increased the size of administrations, creating a modern trajectory of social mobility. Some states such as Egypt engaged in land reform, transferring land away not only from colonial farmers but also from wealthier farmers who benefited from traditional social hierarchies.

Academic scholarship during this period (i.e., 1950s through 1970s) either accepted uncritically development policies and concentrated primarily on implementation and effects or attempted to demonstrate the systemic problems that hindered capital accumulation and income redistribution. Scholars in the former category such as Alex Inkeles (1974, 1983) studied changes in values and the construction of individualism in response to modern institutions and technology. Samuel Huntington (1968) championed the strong state capable of pushing through reforms, particularly those favorable to the West. Scholars in the latter category such as Cardoso and Faletto (1979) and

Peter Evans (1979) showed through their work that alliances among multinational corporations, states, and local bourgeoisie in Latin America created new forms of economic dependency and minimized the role of nonelites in development.

Modernization policies, for the good that they did in expanding social services and building infrastructure, were too costly for states to maintain. They also did not reach large portions of populations and did not encourage, in many cases, competitive economic policies. Furthermore, beginning in the 1980s, Third World countries lost the influence they had gained during the Cold War Era and with it, sources of foreign financial support. The bargaining power of the Group of 77 (the UN coalition, founded in 1964, which represents the interests of developing countries) waned without bipolar conflict. Burdened by insufficient economic resources and pressured by creditor states, countries started to shift away in the 1970s from modernization and import-substitution policies toward more open markets and state retreat from provision of services.

The now hegemonic model of liberal market democracy posits that economic growth results from both integration into the global market economy and international competitiveness in human capital and industry. States have to implement structural reforms that challenge the authority and control they have worked for decades to cultivate. These reforms include financial measures, such as currency devaluation, reduction or elimination of customs duties, and budget cuts. They also entail reforms of the bureaucracy and judicial system and privatization of services.

Scholars of international development likewise have analyzed how different social and political groups respond to reform measures. For instance, the business elites that had benefited from a closed economy may fight more open markets, or unions may react against efforts to liberalize labor laws. Some of these scholars have demonstrated the importance of the state and institutions for promoting business (Evans 1996). Others have condemned neoliberal policies for continuing to benefit political and business elites while engendering suffering among the rest of the population, who remain deprived of adequate services and income (Ferguson 1994; Stiglitz 2002; Crewe and Harrison 1999).

Responding to inadequate progress on social issues under a market-driven model, the United Nations Development Program (UNDP) has emphasized human development and human rights as development strategies alongside economic growth. This move essentially disputes the logic that economic development produces social development. States and domestic and international civil society actors have also responded to mixed records of success of the reform packages by resisting the prioritization of economic reforms at the expense of human development. The consequence has been that even the World Bank and the International Monetary Fund (IMF) now push for institutional reform and support for basic human rights.[12]

Outlining Oxfam America's position on the shift from the emphasis on economic growth to the inclusion of institutional and social concerns, Offenheiser and Holcombe write, "For 50 years we assumed that governments first, and then the market, would provide for basic needs, but each has failed to address the deeper problems of social justice and transform the embedded systems that reproduce poverty."[13] For Oxfam and other NGOs, strategy must take into account this failure by acknowledging the subjective position of the recipients of aid, so long subsumed within specific policies addressing particular issues, from fertility rates to malnutrition. The poor now become actors who can articulate what hinders them in their efforts to improve material conditions and overall quality of life. Development becomes about helping them overcome obstacles to achieving the lives they deserve. This approach to development "assumes that those who are most directly affected know firsthand what institutional obstacles thwart their aspirations and who are the essential actors in deciding what to do about it. Rather than imposing cookie-cutter solutions, this strategy is anchored in the reality of local context."[14]

The basis of such an approach is the 1986 Declaration on the Right to Development. The first two articles state that "the right to development is an inalienable human right by virtue of which every human person and all peoples are entitled to participate in, contribute to, and enjoy economic, social, cultural, and political development, in which all human rights and fundamental freedom can be fully realized" and that "the human person is the central subject of development and should be the active participant and beneficiary of the right to development."[15] At the same time that the declaration spotlights the human facet of development, it criticizes indirectly how states have (not) acted in fulfilling the rights of the subject. The third section of Article Two declares that "states have the right and the duty to formulate appropriate national development policies that aim at the constant improvement of the well-being of the entire population and of all individuals, on the basis of their active, free, and meaningful participation in development and in the fair distribution of the benefits resulting therefrom." Article Three emphasizes that "states have the primary responsibility for the creation of national and international conditions favourable to the realization of the right to development."

The Oxfam article faults states directly for not addressing inadequacies in income distribution or access to services and resources. "Government efforts to address problems are halfhearted or underfunded, or promised funds are diverted into the pockets of urban-based actors who make a profitable career as gatekeepers of foreign-aid programs. The poor are treated as objects of charity who must be satisfied with whatever crumbs drop their way."[16] Their complaint echoes the discourse of good governance, discussed later in this chapter, which represents the latest incarnation of development agency strategies to increase government efficiency while diminishing the prevalence of patronage and corruption.

This strategy may be necessary to ensure implementation of development policies and to work toward greater popular participation in development, but it also reflects the discrepancy inherent within the "development" process. The World Bank, as well as NGOs such as Oxfam that are typically critical of the Bank, pressurize the state which, in turn, must appease the groups that have supported its power for decades, or risk dissent.

Beyond the need for greater accountability and transparency, the Bank and Oxfam may disagree on the role of the state in regard to development. For example, in Oxfam's response to the World Bank's position on loan conditionality, Oxfam presses the bank to remember to respect domestic decisions over development programs rather than insist on control over policy substance. Oxfam writes, "Not imposing policy options means fully respecting domestic political processes, and ensuring the active involvement of parliaments and civil society in scrutinising reform options."[17] The Bank, however, has considered such control part of its overall theme of institutional reform to support market functioning and the promotion of social capital.

The next section explores these tensions between NGOs and institutions. In doing so, the section offers an overview of the multiple actors and viewpoints that frame the field of "development" and influence decision making within Morocco.

THE MULTIPLE VOICES OF GLOBALIZATION AND DEVELOPMENT
The Debates

In his opening remarks at the Shanghai Conference on poverty reduction in May 2004, the then president of the World Bank, James Wolfensohn, remarked, "We start with the recognition that in our world of six billion people, one billion have 80% of the income and five billion have under 20%. We start with the proposition that in the next 25 years, two billion more people will come onto our planet, and all but 50 million will go to developing countries."[18]

Wolfensohn also said that he remembered the 2003 G-8 Summit in Evian, where President Luiz Inácio Lula da Silva of Brazil mentioned that perhaps the following year the G-8 leaders should consist of those governing the countries that represent the majority of the world's population. "[Lula] pointed to this new balance that is needed in our world. He pointed to the fact that today there is an imbalance and that we have a challenge of poverty before us which has been identified in the Millennium Development Goals."[19]

Although many NGOs might argue that the World Bank's policies have exacerbated the imbalance, they would probably agree that the mission of development is to fight poverty and inequality. Global Exchange, a human rights organization based in San Francisco, calls the World Bank and the IMF "the world's biggest loan sharks,"[20] while championing the same objective of improving the welfare of the world's most vulnerable populations.

Their primary difference with the World Bank as well as the IMF lies in perceived solutions. Whereas the World Bank and the IMF provide loans in order to implement prescribed social and economic policies, an organization such as Jubilee Research, the heir to Jubilee 2000, has worked to forgive the debts owed by low-income countries and to grant those same countries more power over policy decision making. Furthermore, whereas the World Bank and the IMF promote integration into the world market as a solution to poverty, most alternative NGOs view global market capitalism as a catalyst for further inequality.

For the World Bank and IMF, free trade agreements function as pillars in developing a global structure for market capitalism. Western governments sign trade agreements in order to access new markets and, ostensibly, to promote domestic growth in the participating countries. Governments in developing countries likewise hope to benefit from both the possibility of exporting to wealthy markets and from an influx of foreign investment. Indirectly, but critically, increased exports and investment should help to reduce poverty and unemployment and promote a higher standard of living.

In a column praising reduction in extreme poverty in East and South Asia, *New York Times* columnist David Brooks extols the virtues of free trade:

What explains all this good news? The short answer is this thing we call *globalization*. Over the past decades, many nations have undertaken structural reforms to lower trade barriers, shore up property rights, and free economic activity. International trade is surging. The poor nations that opened themselves up to trade, investment, and those evil multinational corporations saw the sharpest poverty declines. Write this on your forehead: free trade reduces world suffering.[21]

Brooks goes on to say that at current growth rates, the UN will attain its Millennium Development Goal of cutting extreme poverty, or living on less than $1 a day, by 2015.

Brooks is delivering for a popular audience the arguments championed by international financial institutions, Western governments and, some, though certainly not all, development analysts (see, notably, Jagdish Bhagwati 2005). Morocco, despite empirical ambiguity about that benefit, the rhetoric that frames free trade agreements in Morocco has continued to promise great rewards. After the ratification of the free trade agreement with Morocco in 2004, U.S. Trade Representative Robert B. Zoellik exclaimed, "Free trade is on offense, and the cause of open markets is now advancing on all fronts … Working together with Congress, the administration is putting Trade Promotion Authority to good use, and America is at the forefront of the global move toward expanded trade."[22] For his part, Prime Minister Driss Jettou envisioned that the benefits of free trade would extend to the wider population, that investment would "create employment opportunities, particularly for our qualified youth."[23]

Opponents of unregulated free trade, and of economic globalization in general, reverse these arguments to equate market reform measures and greater insertion into the world economy with the breakdown of traditional social units such as communities and families. They criticize the universalism characteristic of aid programs and the lack of specific knowledge that officials at large institutions and government aid agencies have of low-income countries. Prominent development scholar Jeffrey Sachs has faulted the Bush administration for exactly this absence and its tragic consequences:

> Whether I look at the National Security Council, the Treasury, the Council of Economic Advisers, the United States Agency for International Development, or the relevant congressional committees, I see woefully few individuals with expertise about the low-income world. This is too bad, because the low-income world (roughly, those who live and die on less than $2 per day) constitutes 40% of humanity—and most of the places where American troops have fought and died in recent decades.[24]

Grassroots organizations, whether based in the West or developing countries, use this deficiency in local awareness to push for a role in decision making over development policy. At the heart of their efforts, however, lies another core set of beliefs about world order. This world order may look different from that disseminated by the World Bank, but it represents a vision just as well, one that envisages a larger role for global social institutions, for treaties on labor, the environment, taxation, and human rights, and for general prioritization of furthering human capability over economic growth and materialism.

Edward Goldsmith, founder of the journal *The Ecologist*, summarizes the most prominent alternative position on global market integration when he writes, "Instead of seeking to create a single global economy, controlled by vast and ever less controllable transnational corporations, we should instead seek to create a diversity of loosely linked, community-based economies managed by much smaller companies and catering above all (but not exclusively) to local or regional markets."[25]

Critics of free trade believe not only in localized economic strategies but also in fair trade consumerism, environmental awareness, and respect for human rights. They likewise support moving away from "fast" capitalism, or deregulation of financial flows, to prioritization of social and environmental needs. Some groups, such as France-based ATTAC, fight for the implementation of the Tobin Tax, a tax on currency transactions, in order to raise funds for global antipoverty, environmental, and health programs.

The Moroccan chapter of ATTAC has adopted all of those positions, declaring at the 2002 Bouznika (Morocco) Social Raram that "we refuse that the politics and the choices destructive of our society are carried out, and that commercial and economic accords are concluded in our name and without our knowledge of them." (ATTAC–MAROC. 2002. "Contre la modialisation libérale." No. Special, p. 3.)

In the e-discussion on globalization and poverty launched by the World Bank after the antiglobalization protests of the late 1990s—the largest of its kind for the Bank, with over 5,000 participants—academics and activists proffered four broad critiques of the development approach of international financial institutions. First, they argued that trade and market liberalization increase poverty and inequality through multiple factors:

the concentration of capital, the flight of capital, the loss of domestic industry, increased competition for small businesses, and the production of unstable, low-quality jobs. Second, they encouraged local and pluralistic development experiments. Third, they supported a nonquantitative measurement of change that replaces the Westernized, anonymous economistic language of development by taking into account the experience of much of the population directly affected by reforms. The fourth argument, primarily put forth by academics, called for more transparency in supranational institutions as well as controls over capital and the prioritization of social development.

These critics, as well as supporters of market capitalism as "development," center their positions on ideological justification as much as on concrete evidence. Their primary beliefs regarding how the world should be ordered reflect their relative claims to power: an imposing institution dedicated to a universal strategy of development, of global market integration, juxtaposed with often politically marginal intellectual elites and tens of thousands of dedicated grassroots activists. A correspondent from the Indian town of Tiruchirapally, Tamil Nadu, complained in a World Bank e-discussion on poverty that in India, a much-hailed national budget presented in 2001 liberalized regulations at the expense of local small businesses. The correspondent wrote,

> The former finance minister's dream budget was hailed all over the world except by the small-scale industries in India. The reason is very simple—the introduction of MAT [minimum alternate tax] tax made NEPC [a company funding alternative energy technologies] shut down its wind energy project. In Tiruchira-pally alone over 100 manufacturing firms which were working

on this closed down, leading to massive unemployment, and the owners of these small-scale fabrication firms ended up with what they considered massive bank loans. Successive governments have done nothing to revive the project despite requests. So apart from the unemployment and debt, production of electricity from wind energy was ended.

According to this correspondent, the Indian government should have ensured the maintenance of a project that contributed to positive environmental policy and assisted both the company, NEPC, and related small businesses, which promote economic growth and social mobility.

In contrast, a correspondent working for the World Bank argued, "I don't know the specifics of the NEPC wind energy project, but the example you cite has a lesson, namely that subsidy-driven projects are not sustainable, unless the subsidy is linked to performance in such a way as to reward cost reduction (and hence, less need for subsidy over time)...." He ends his message with the comment:

What did these heavily subsidized investments yield in the end? Who knows or cares? In essence, we did the same thing to wind that we did—and are still doing—to nuclear: massive government support for the sake of feeling good about ourselves and showing off to the outside world. (Who cares if the cost of showing this technological prowess is that millions go hungry and unclothed?) It's sad what happened to workers in Tiruchirapally, but surely there were, and are, better uses of government money than supporting businesses that can't otherwise survive in the marketplace?

For the employee of the World Bank, rules of the market and not government intervention or social or environmental concerns should determine economic survival. Those who survive the market demonstrate a sustainability that ultimately fights poverty more effectively than the subsidized, interventionist policies favored by critics of the World Bank.

Ideological differences concern political as well as economic transformation. The United States Agency for International Development (USAID), following the logic that democratization is good for national interests, promotes specific programs relating to democracy as part of its mission. Officially, the Democracy and Governance office "identifies the democracy and governance needs of a country, assesses the country's commitment to democratic governance, and develops country-specific plans to address the country's needs or to enhance its contribution to the resolution of regional or global problems."[26] These plans can include encouraging elections, expanding civil society, and making the judiciary more transparent and effective.

In Morocco, the goals are potentially more explicit and more ambitious, namely, to provide "overarchiving U.S. foreign policy interests promoting regional stability, economic development, and democratic values, and combating international terrorism." USAID's program in Morocco, as in countries like Egypt, include democracy training for local political parties and members of parliament. (www.usaid.gov/policy/budget/cbj2005/ane/ma.html). As with World Bank programs, economic self-interest is never far off. A balanced judiciary system would assure foreign companies that their investments would be protected by laws, rather than be subject to discrimination by local politicians or business competitors.

It is this economic interest, as well as desire for political hegemony, that international and local grassroots organizations often resist. Their own strategy, articulated through international social movements and conferences, like that in Porto Alegre, Brazil, in 2002, calls for more autonomy for local populations through global protection from economic exploitation and political repression. Individual organizations such as Global Exchange assist human rights activists in countries where military repression and largely American policies have allowed for abuses, particularly to poor populations. Sweatshop Watch, also based in San Francisco, fights for improved working conditions and better pay for garment industry workers both in the United States and around the world. The organization, working with the labor union UNITE, sponsors boycotts of companies—most notoriously NIKE but also Donna Karan—that Sweatshop Watch claims use production units that prevent bathroom access, limit overtime pay, and demand long hours.

In other words, the NGOs that participate in these movements pursue fashioning universal laws that evolve from the specifics of a location versus establishing universal rules in order to make the local more global. Mark Ritchie, the president of the Institute for Agriculture and Trade Policy, refers to this process as "trade managed for sustainability." He explains that "trade, like all other commerce, needs to be managed for sustainability—fair prices, profits, and wages for everyone making a contribution to the final product."[27] Ritchie also expresses optimism that the international social movements that have maintained pressure on trade and aid policies will achieve goals of international systems of protection.

There is a very high level of global understanding, solidarity, and active cooperation at present—north, south, east and west … We now have the opportunity to draw on the lessons, experiences, encouragement, strengths, wisdom, information, and resources of colleagues from all over the planet at literally a moment's notice, and we are increasingly doing this to advance sustainable human development and human rights. I have never felt so encouraged by this aspect in my entire political life.[28]

According to him, this network offers a challenge to governments ruled "by elites for elites," a system he insists inevitably ignites or exaggerates conflict.

The Players in the Field

Between the IFIs such as the World Bank and IMF and organizations that participate in the World Social Forum, such as ATTAC, a number of large NGOs operate such as Save the Children and World Vision, as well as a growing number of private consultancy firms such as John Snow Inc. (health issues) and Chemonics International Inc. (economic development as well as health and environmental issues) operate (see Stubbs 2003). All these particular organizations and consultancy firms, with the exception of World Vision (an explicitly Christian organization), have managed projects in Morocco.

Consulting firms and international NGOs partner or fund in Morocco with some of the tens of thousands of associations that have appeared since the liberalization of laws in the early 1990s. For example, the German Freidrich Ebert Foundation sponsors seminars on topics such as civil society, journalism, and politicization of youth in coordination with both government agencies and organizations such as

Association Marocaine des Droits Humains. The Japanese and French governments work with the government on large-scale projects—in Japan's case, water supply in rural areas—and provide small grants to local associations for specific programs. As we discuss in Chapter 2, some of the more elite or urban associations do work with more political organizations based in Europe and the United States. Both ATTAC and the Germany-based Transparency International, which acts to reduce corruption, have chapters in Morocco.[29] However, most of the thousands of associations concentrate on very local economic and social issues, often in rural rather than urban areas.

Large international organizations such as Catholic Relief Services (CRS), Save the Children, and the Grameen Bank (organizations that have offices throughout the world) often work at an intermediary level between these grassroots organizations and the IFIs such as the IMF. They integrate ideas that support dominant global policy agendas, for instance, fostering civil society, female education, and "good" governance, with specific missions. Save the Children works with children and families and therefore concentrates on establishing programs concerning maternal and child health and education. Similar to the World Bank or UN agencies, CRS sponsors programs that promote state–NGO partnerships, rural infrastructure, and female literacy. In Morocco, as part of its goal to encourage the growth of civil society, CRS founded AMSED, a national umbrella organization for community development organizations. AMSED's responsibilities range from training in management to fostering programs in areas such as HIV/AIDS education and microcredit—the granting of small loans to those living in poverty who would not meet minimum requirements for standard credit.

NGOs of all sizes and IFIs achieve influence through the dissemination of generic models and concepts and successful local program impact. For instance, *good governance* refers to laws and regulations against corruption, an independent and effective judicial system, an independent press to monitor government actions, and oversight procedures within governmental institutions themselves. Each of these steps should counter political repression and concentration of power. At the same time, organizations and institutions like the World Bank encourage the growth of *civil society*, meaning an increase in the number and activities of NGOs, faith-based institutions, community groups, labor unions, charities, and so on, that engage in political discussion and, more importantly, social work.[30]

In fact, the language used to describe programs—as well as the design of projects—can overlap between IFIs and large, international NGOs. When the UN announced the Millennium Development Goals in 2000, one goal in the resolution was to "promote gender equality and the empowerment of women as effective ways to combat poverty, hunger, and disease and to stimulate development that is truly sustainable."[31] CRS claims as one of its goals in Morocco, "[to] place a special emphasis on women, working to increase their voice in the decision-making process in project activities. The literacy program focuses on women's empowerment, education and literacy training, and income-generating activities."[32]

Both financial institutions and large NGOs construct programs according to expected outcomes. Many of these programs now follow the direction of the Millennium Development Goals. The Millennium Development Goals represent global targets for 2015 (with a few exceptions). The objectives range from cutting extreme poverty in half (see above) to reducing child mortality by two-thirds and the number of women who die in childbirth by three-fourths. Agencies such as UNDP and their local partners assume responsibility for translating these broad, universal aspirations for social change into projects on the ground. UNDP in turn measures the success of a program by its capacity to advance toward the universal objectives.

Morocco has made better progress than many other countries in achieving the 2005 goal of ending gender disparity in school enrollment, leaping from a rate of female enrollment in rural areas of 44.6% in 1997–1998 to 82.2% in 2002–2003. This percentage still fails to meet the Millennium Development Goal of 2015 of equal benefit from education, as the primary school completion rate for girls hovers around ten percentage points lower than that for boys, or 76% compared to 85%.[33]

The method of implementing and evaluating a model by turning full circle—from idea to practice back to idea—characterizes evaluation practice in international organizations. For instance, the United Nations agency International Fund for Agricultural Development implemented a project of organizing cooperatives for farmers in the eastern part of Morocco. The project took the general and currently popular idea of a cooperative and applied it as a solution to problems of range management, namely land degradation versus sustainability. Emulating goals disseminated by other UN agencies, some of the project's components included offering credit to small farmers and supporting women's activities as well as improving the quality of land and water supply.

The evaluation of the project involves logically analyzing, among other results, the effectiveness of the cooperatives.

These cooperatives had been built upon existing ethnic and social structures and ended up replicating hierarchies to the disadvantage, perhaps, of smaller farmers. The evaluation report reads, "As to the viability of the formula, one must find out if cooperative rules will, with the passage of time, be able to impose novel rules of operation on social structures that have been made largely unbending by traditional practice."[34] In other words, the report asks if the social structure can be altered to fit the idea, rather than questioning the cooperative as a concept or practice suitable for the social structure. The evaluation envisages social alteration to accommodate the generic solution.

In practice, the application of a universal model involves mutation into disparate forms, depending on the participation of local NGOs and leaders. The question becomes, how do these actors use the ideas and funds provided for development projects to transform the space in which they live and work, both in terms of material welfare and political and social identity? Responsibility for social services can facilitate state penetration but in a way considerably different from that of the centralized, authoritarian structure of the modern nation-state. Social services can also take on forms not envisaged or even desired by their founders, whether in an international NGO or the World Bank.

Microcredit, almost universal now as a technique of providing capital for the poor, originated with the Grameen Bank, an organization founded in Bangladesh but now existing worldwide. AMSED created the first microcredit program in Morocco in 1993. Since then, a number of microcredit organizations have appeared, ranging from sizable foundations such as Al Amana[35] and Zakoura that were started by wealthy philanthropists to small associations located in small cities and towns such as

Khenifra (central Morocco). As with any program, particularly those implemented from a general model in a particular setting, microcredit projects in Morocco have met with mixed success. On the one hand, three national banks (Credit Agricole, Banque Populaire, and BMCE) have decided that the public relations and social value of microcredit merits investment, and have thus created microcredit programs. Loan reimbursement rates are also very high. For instance, for Zakoura, the reimbursement rate is 99.79% for over a million loans totaling an amount of more than 100 million MAD. (MAD is the abbreviation for the Moroccan dirham.)

On the other hand, the cost of loans for the recipient can be very high. The associations providing the loans borrow money from banks and then charge a higher rate of interest in order to fund their services, passing the cost on through interest rates to the borrowers. Finally, the types of businesses that borrowers start often lack sustainability because of their small size, insufficient capital, and inexperience in the market. The American director of a microcredit organization based in Rabat told one of the authors that "all of the loans have been a success." Yet, in the same conversation, he admitted that the loan recipients lacked the experience to evaluate the market and create a niche for themselves.[36]

Microcredit, as a loan program to establish microbusinesses, clearly reflects the preeminence of neoliberalism and market capitalism as trajectories of development. We can situate microcredit as a model within a field of development dominated by a hegemonic ideology of economic growth with global targets for improved social indicators. NGOs and international financial institutions such as the IMF confront or cooperate with each other over models such as microcredit as well as larger issues of market reform. They each use their political allies and mobilization tactics, from interviews in the media to public speeches to Internet campaigns, to persuade other international NGOs and local actors in developing countries to believe in their approaches. Ultimately, though, they all target national policy and the state, whether indirectly through supporting minority communities in deprived areas or disaffected populations (such as ATTAC's work with unemployed university graduates in Morocco) or directly, through meetings with high-level government officials (such as the IMF's regular country assessments known as Article IV Consultations).

WHAT ROLE FOR THE STATE?

Market and political liberalization have thrown into question the natural interests of states in developing countries and the elements that have traditionally framed their authority. States in newly independent countries in the last century needed to consolidate power and demonstrate legitimacy within the international system of nation-states. Today, states must determine the most effective trajectory toward economic growth and political stability in addition to managing their financial obligations to the World Bank and IMF.[37] First, the state must engage in reforms that alienate nonelite populations dependent upon public sector services and employment and elite groups protected by tariffs, monopolies, and patronage. Second, the state must turn upon itself, to trim bureaucracies, fight corruption and waste and, ultimately, reorganize the structure of its own power. More fundamentally, with free trade and shrinking capacity to intervene in the construction of a "nation," the state is losing the purpose it had in the past.

In Morocco's case, recent trends have undermined alliances and tacit agreements that have helped to buttress the state's authority since independence. Since then, the state has protected the local bourgeoisie by regulating trade and domestic production and built a middle class through expansion of the public administration and education. The state has also allowed practices such as corruption, which may have not only hindered business development and discriminated against those with lower income and few social contacts but also given individuals ranging from petty bureaucrats to high-level military officials greater income and, thus, more contentment.

Yet, for the World Bank, foreign government aid agencies, and organizations such as Save the Children, these practices have undermined economic growth and social equity. A World Bank report, lamenting slow progress on "good governance" in the Middle East and North Africa, blames inefficiency and corruption for ills ranging from unemployment to low rates of private investment to inadequate social services. The report offers as evidence the correlation between the facts that countries in the region maintain the least amount of data on governance in the world, power is concentrated in the executive, and economic productivity has declined over the past thirty years.[38] The report states, "Many factors contribute to the region's disappointing economic performance, with weak governance at the origin of many. Governance helps determine policy formulation and implementation which, in turn, determine whether or not there is a sound, attractive business environment for investment and production."[39] The report specifically cites a survey of businesses in Morocco, in which half the respondents testified that they used intermediaries or hired full-time staff to manage bureaucratic procedures.[40]

The solution to these problems is public sector reform, namely decentralization of decision making (enabling greater flexibility), better training and fewer employees, and simpler, more transparent bureaucratic procedures. The World Bank 2001 country assessment for Morocco calls for sustained and considerable changes to centralized decision making, the organization of the civil service, and the basic tasks of central government:

> The state in Morocco remains heavily involved in the provision of goods and services, well beyond what would be justified by objectives of regulation or redistribution, and in excess of its capacity to provide strategic leadership. Public sector management has been traditionally characterized by excessive centralization and a culture of administrative formalism, which has gradually degraded into poor accountability and pervasive petty controls. Not only does this affect efficiency, but it also creates opportunities for mismanagement and corruption, cited by private enterprises as one of the main obstacles to their development. The government has placed the issues of transparency and corruption squarely on its agenda.

For its own part, the government has trimmed bureaucratic procedures to establish businesses by creating the Centres Regionaux d'Investissement (CRI), which in turn facilitates acquisition of land and makes it easier to obtain proper paperwork. However, simplifying regulation of business at a local level has not done away with all obstacles. Business owners must still ensure that the land they purchase has proper infrastructure, with electricity, water supply, and so on, which depends on the municipality.

Students of development have investigated for decades how changing the way in which institutions and regulations operate—such as establishing the CRI but not pushing for corresponding support through municipal governments—affects accumulation of capital. Certainly, we could apply some of their theoretical arguments, which include game theory as well as more sociological analyses of social relations between political and economic actors, to Morocco. Based on the lessons of East Asian countries and India, we could argue that the Moroccan state has not built alliances with capital that have encouraged creativity and competitiveness through a combination of regulation and incentives (Evans 1996). The economic dependence upon often languishing textile and tourist industries shows that the state has not encouraged sufficient diversification into industries that require higher skill levels and also yield greater income. In addition, the state has not adequately prepared the existing industries for free trade.

We could also argue that actors in Morocco critical to generating economic activity have not perceived enough benefit to change their behavior (see Bates et al. 1998), for instance, by taking risks by investing in a new industry (rather than, for instance, in real estate and construction). At the risk of seeming overly simplistic, we could suggest that the black market economy in Morocco, which consists of trade in hashish and contraband, reflects the self-interest and game playing of multiple actors. Much of the income generated in the northern part of the country comes from producing and trafficking cannabis or trading in black market goods. The sale of hashish may yield anywhere from 2 to 20 billion dollars a year. The networks that smuggle contraband and run hashish from Morocco to Europe, at least in the North, are increasingly interpenetrated with radical Islamist groups and lobbies close to state functionaries, from gendarmes to customs officials.

If the state elects to control the black market and drug economies, then the King and government officials will encounter opposition not only from powerful political figures but also from small farmers and low-income families that depend on illicit trade for survival.[41] However, the state must seek to limit the global and local reach of both the drug cartels connected to industry and politics and the Islamists in order to prevent rival systems of power. European governments support the state in this situation, in that they have invested in cannabis replacement schemes, yet they also do not want to contribute to the generation of greater poverty and alienation in the North. These replacement schemes—to grow saffron, for instance—have not significantly reduced production. In the end, whatever action the state, producers, traffickers, and European governments take regarding the black market, they do so based on their separate political and economic interests as well the perceived loss or gain of the other actors involved.

Could we, though, use these kinds of explanations to understand why Morocco has not made more progress on social issues and in promoting human capital? Critics of the World Bank could say that in following the recommendations of the World Bank and IMF, the state has limited its options for public investment. In fact, the state must negotiate factors too complex for straightforward, singular strategies. As we discuss in the rest of the book, the state must demonstrate both its existential and practical value to the citizens of Morocco in order to compete with and manage Islamist and democratic political movements. The technical approach of the World

Bank or even a critic such as Oxfam ideally has more than a material impact, in that the beneficiaries have greater access to basic services and income and, thus, greater freedom as human beings. However, for policy reform to have more than an assumed, nonmaterial impact, the state must redefine its moral and material purpose for the individual and the collectivity. This collectivity cannot merely replicate the reproduction of inequality of the past and present, as market reforms have done, but move beyond to a reflection on reconnecting social groups and restoring social mobility.

FOLLOWING DEVELOPMENT POLICY
Morocco and Modernization

During the decades after independence, similar to many other countries in the Third World, the regime in Morocco concentrated on domestic, economic, and social policy. Through five-year plans, the regime controlled economic diversification, modernization, and social investment. The regime adopted the dominant logic of development at the time and prioritized economic growth over social development, believing only economic growth would bring about improvement in social indicators. Yet, simply following the basic rules of development did not lead to sustained economic growth or sufficient human capital and industrialization to become competitive in the world economy. For reasons ranging from continued concentration of income to inadequate progress on social issues such as education, Morocco at the end of this period found itself heavily in debt and ill-prepared for the next development ideology—integration into the global market.

During this post-Independence period—from the 1950s until the early 1970s—the state extended its control over formerly French-owned major industries such as phosphates and protected the domestic market and industry from foreign competition. However, the state did not go beyond protection to promote local industry's competitiveness in the global market. In addition, the state did not foster economic linkages within Morocco, meaning that industry still imported critical goods.

By the 1970s, economic growth picked up (average annual GDP growth of 7.3%), primarily owing to receipts from the mining and export of phosphates.[42] After two failed coups d'etat against the King in the early 1970s, the regime used growth to invest in the rise of a loyal middle class that would equate its interests with state legitimacy. At the same time, though, large sections of the population, especially the rural poor, did not benefit from social policy.

For example, households earning the lowest 40% of income dropped in percentage of total consumption from 18% in 1959–1960 to 12% in 1970–1971. The number of individuals living in poverty rose by 1.3 million between 1960 and 1977, including a million in rural areas. The rural poor did not benefit from continual growth in agriculture during the 1960s (3.7% average annual increase in value added from 1960–1971) and suffered from the lack of growth in the 1970s. This absence of benefit was because, among other factors, of the structure of land ownership, population growth, and vastly inadequate public investment into infrastructure in rural areas. The social consequence was rural–urban migration, contributing to a trend that has prevailed throughout the developing world.[43]

The failure to diminish poverty in general and specifically in rural areas, along with the slow expansion of education and women's rights, left the state vulnerable to external criticism. At the same time, declining economic growth and burgeoning debt increased obligation to foreign creditors. Medium- and long-term debt increased from 19.6% of GDP in 1975 to 85% in 1983.[44] In response, the Club of Paris, a group of creditor nations, negotiated in the early 1980s the beginning of structural adjustment and market reform and formally ended the post-Independence strategy of domestic protection.

As with modernization and nationalization, Morocco proved itself to be a good student of the new development ideology. Still, Morocco found itself at the center of debate about why successful implementation of reform did not translate into accelerated economic growth, economic diversification, or improved social indicators. Consequently, from the mid-1990s onward, social progress, institutional reform, and the process of making policy decisions became far more significant in official and civic debate.

Morocco and Market Reform

Morocco has complied with much of the World Bank and IMF's conditions for restructuring over the past several decades. This compliance has intially come out of financial necessity from the obligation to confront large budget deficits. However, with the successful implementation of structural adjustment policies, Morocco has gained credibility and, thus, bargaining power regarding World Bank and IMF recommendations.

Policy decisions over the past several decades reflect Morocco's negotiations with IFIs and the resulting change in focus from economic difficulties to administrative and social problems. During the 1980s, the government devalued the currency, reduced customs fees, eliminated export taxes and import restrictions, and increased control over public expenditures. By the mid-1990s, budget deficits had decreased to approximately 3 to 4% of GDP.[45]

From the 1990s until today, Morocco has proceeded perhaps too slowly for its creditors on privatization, but the state has remained faithful to the basic tenets of market liberalization. The state has reformed the labor code and the judicial system and signed multiple free trade agreements. The state has also devalued the dirham, pegging it to a basket of currencies, and reduced public debt relative to GDP.[46] Yet, these measures, as well as the financial ones that preceded them during the 1980s, have not succeeded in accelerating economic growth. The annual growth rate in GDP per capita from 1990–2003 was 1%.[47] The only exceptions have resulted from profitable years in agriculture and the sale of state-owned enterprises, for instance, Maroc Telecom. Overall, the GDP grew at an annual average of 1.8% during the 1990s.[48] Since then, growth has increased, but not in a consistent fashion. In 2001, real GDP grew by 6.3%, decreased to 3.2% in 2002, and grew again to 5.5% in 2003[49] (with a preliminary estimate of 4.2% in 2004 and a projection of 1.0% in 2005).

As in many other countries engaged in market reform, budget cuts, debt reimbursement, currency devaluation, and insufficient economic growth have negatively affected living conditions, particularly for the urban poor dependent on public services. Public sector expenditure decreased from 34.4 to 22.2% of GDP from 1982–1986. In the second half of the 1980s, public sector expenditure began to increase again, to

25.9% in 1991. This level, or several percentage points lower, has continued until the present.[50]

The inconsistency level of economic growth and the decline of public intervention in Morocco have corresponded with high rates of poverty and unemployment. The World Bank has estimated that Morocco needs regular growth of 6% or more to reduce overall urban unemployment rates, which hover around 18%, as well as the 30% unemployment among educated young Moroccans (past primary level).[51] Morocco would need similar rates of growth—5% or more a year—to decrease significantly the number of poor.[52]

Poverty has remained a consistent problem for over a decade, reaching 19% by 1999, although the poverty rate supposedly has dropped slightly in recent years to 18% in urban areas and 15% in rural areas (IMF Article IV Consultation 2005. p. 5). Overall, approximately 5.3 million individuals live in poverty in Morocco, out of a population of more than 30 million.[53] Continuing high rates of poverty in urban areas reflect a lack of growth in industry, shrinking public services, and rural-urban migration. In rural areas, the poor depend on favorable weather, availability of basic services such as electricity and roads, and investment into agriculture. Until the past decade, state policy did not coordinate rural economic and social policy and, thus, rural poverty continued as an offshoot, positively or negatively, of economic growth.

In response, the government, with World Bank funding, created the Agency for Social Development. The King also recently launched the National Initiative for Human Development to combat poverty in 250 urban and rural communities through developing infrastructure, social services, and support for income generation. The initiative, funded in part by local governments and the state and in part by international actors such as UNICEF and the French government, should total 10 billion MAD over the years 2006–2010. Recent policies targeting provision of electricity (PERG), roads (PNCRR), water (PAGER), and regional development (BAJ1) have been successful but also reveal how much of the population, urban and rural, still requires intervention. A World Bank evaluation of BAJ1 concludes that "although the BAJ is concluded to successfully target the poor ... The BAJ is implemented in provinces that cover at most only half of the country's rural poor and under 40% of its total poor."[54]

Morocco has also accelerated integration into the global market in order to jump-start economic growth. However, free trade agreements with the European Union and the United States until now have not yielded the increase in exports and foreign currency necessary to push Morocco into a better economic position. In 2004, foreign investments and loans declined by 71.6% to 6 billion MAD, essentially because of an absence of major privatization initiatives.[55]

Despite uneven results from market reform, the World Bank has remained optimistic about the state's sincerity in improving social indicators and continuing to liberalize the political system. Although the World Bank has repeatedly criticized Morocco's relatively poor progress on social issues, the Bank does commend overall achievement:

> During the last 30 years, Morocco has embarked on one of the most successful programs of human development and political liberalization in the Middle East and North Africa (MENA) region. Since the 1970s, gross national income per person has more than doubled from $550 to $1,190. The

average life expectancy has increased from 55 in 1970 to 68 in 2001. During the same period, the average number of births per woman has seen a dramatic decline from 6.3 to 2.8, whereas the number of children dying before age one has dropped from 115 to 39 (per 1,000 live births). Substantial educational improvements during the past 30 years include a primary school enrollment increase from 47 to 90% by 2002.[56]

Such political support for improvement in social conditions and liberalization are in turn altering the landscape of political mobilization and conflict. As we discuss in Chapter 3, both actions on the part of the state have allowed for the emergence of popular mobilization and have made constraints on political freedom a topic of public debate.

Political Liberalization

In the early 1990s, faced with budgetary constraints, weak economic growth, and growing social unrest, King Hassan II undertook limited measures of political liberalization. Among his first steps, as we discuss further in Chapter 2, were the release of political prisoners held since the 1960s and 1970s and the closure of a desert prison, named Tazmamart, which had become infamous for human rights abuses. By the mid-1990s, he had also convinced opposition leaders to support institutional reforms and, to eventually, participate in elections that brought them into power. King Hassan II created a bicameral legislature in which the lower house would be directly elected and the upper house elected through municipal councils, union representatives, and business councils.

This system, however—which retained the monarch's ability to choose candidates for certain ministerial positions and pushed through economic reforms without dissent—left the power of the King intact. Moreover, when in 1997, the opposition government led by the Socialist party (USFP), came to office, it assumed the obligation of managing the social problems engendered by budget cuts and weak economic growth.[57]

The seeming cynicism behind initial political reform does not belie the fact that political liberalization has continued, albeit haltingly. The opposition still governs, political exiles (including the leader of the USFP in 1997, Abderrahman Youssufi) have returned to Morocco, and human rights has become a critical topic of discussion. The range of media available has exploded, and despite censorship, magazines and newspapers continue to test the boundaries of critical analysis and investigative reporting. Perhaps more importantly, as we explain in more detail in Chapter 2, the emergence of new political movements, including the Islamists, and pervasive alienation among the young have made liberalization potentially irreversible.

The response from creditors and from international organizations, such as Human Rights Watch and Amnesty International, to political liberalization in Morocco has generally been praiseworthy. Amnesty International reported in 1999 that the human rights situation in Morocco had indeed improved dramatically. The state repealed the 1935 *dahir* (decree) used to arrest prisoners of conscience; limited the time of solitary detention; released hundreds of political prisoners and prisoners who had "disappeared;" and commuted almost 200 death sentences. The state also created the Consultative Council for

Human Rights in 1990 and the Ministry for Human Rights in 1993 and ratified the Convention against Torture in 1993.

In the course of the past few years, a number of legislative and institutional measures have been taken that have resulted in significant improvements in the human rights situation in Morocco. These developments have meant that Moroccan civil society has been able to enjoy appreciably higher levels of freedom in recent years. Political parties, independent media bodies, and NGOs are widespread and active in Morocco. Criticizing the monarchy and expressing views in favor of the independence of Western Sahara remain taboo, but Moroccans are increasingly able to engage in opposition activities, criticize government policies, and conduct open debates on human rights issues. The reservations of Amnesty International regarding Morocco today are related to detainees and reports of abuse in Western Sahara.

Improving the position of women has also assumed importance. Politically, the King has allowed for a change to the family code and mandated a quota for female members of parliament. The state has pushed for literacy and maternal health programs in order to reduce high rates of illiteracy, particularly in rural areas, and high rates of maternal mortality. In 2003, the overall female literacy rate was 38.3%.[58] A 2001 study conducted by the Ministry of Health found that 227 per 100,000 women in Morocco die of causes related to pregnancy or childbirth. These figures represent little progress over the last study, conducted in 1997, which found that 228 per 100,000 women die of similar causes.[59]

The difficulty facing policy directives such as those of poverty reduction and improving the position of women is that they deal with only parts of general inequality. Without confronting the conditions that produce unequal distribution of resources, the state cannot overcome overwhelming rates of poverty and unemployment, and the social restlessness that accompanies them. Reduction of inequality within Morocco seems to remain elusive, beyond the reach of current development policy. According to a government study in 2003 on regional competitiveness, almost 40% of wealth is concentrated in 1% of territory. More than 77% of territory contributes around 10% of value added to the economy. The Kenitra–El Jadida corridor, which includes Casablanca and Rabat, accounts for 61% of the urban population, 80% of permanent jobs, and 53% of tourism.[60] These figures on regional disparity are compounded by the disparity between socioeconomic groups, as we discuss in Chapter 4. In other words, the King must engage in social intervention and political maneuvering far more seriously if he and the state are to combat truly the question of poverty or borderline poverty.

CONCLUSION

Despite its relatively weak rank in social indicators and the persistence of poverty, in the language and evaluation techniques of international organizations, Morocco has "moved forward." Yet, process of economic and political liberalization have been far more complex and contradictory than following the recommendations of the World Bank. We have to ask what a country such as Morocco, and economic liberalization, can do when faced with multiple political, economic, and social problems, in part generated by those very recommendations. For instance, political liberalization has not been able to stem the rise of Islamism and market liberalization has not been able to reduce unemployment, particularly in urban areas, or reduce poverty. In fact, both have corresponded with an increase in political radicalism and in social and economic marginalization.

All the discussions of greater human rights, trade agreements to promote exports, and the surge in the number of community development organizations—to about 30,000 by the end of the 1990s—have not necessarily put Morocco on a rapid path to better social welfare, but that fast track is what Morocco needs. In addition, as we will discuss in Chapter 4, market liberalization has provoked social reorganization, introducing new patterns of social marginality and providing space for new identity formation that rejects national identity and a national project of development. Foreseeing difficult futures, unemployed young men and women look to migration for aspirations to a better future, rather than local possibilities. The social structure that supported the monarch for forty years after independence, for better or for worse, is eroding. We must then ask: to what effect for Morocco?

INTRODUCTION

During the 1980s, faced with pressure from the Club of Paris, a group of creditor nations, and international organizations such as Amnesty International, King Hassan II began to release political prisoners arrested in the 1960s and 1970s. Most of these prisoners had either participated in Marxist movements or been associated with two failed *coups d'etat* in the early 1970s. In 1991, Hassan shut down the notorious Tazmamart prison, where he had confined the officers arrested in connection with the coups d'etat. Ahmed Marzouki, in his chronicle of the eighteen years he spent at Tazmamart, described prison cells three meters long and two and a half meters wide with a slab of cement for a bed. Along one wall lay the toilet and holes for air passage. Deprived of light and companionship, he and his fellow prisoners were permitted to leave these cells approximately thirty times during their imprisonment. Marzouki recalls his surroundings:

> The furnishings of the cell were neither spartan nor rudimentary, they were reduced to the simplest expression. A plate, a carafe, and a pitcher of water, which could contain five liters, were set down near what we could call pompously, or derisively, our toilets. The pitcher was filled once a day and this small amount had to serve for drinking, bathing, and cleaning our

clothes. Is it necessary to point out that in eighteen years and thirty-nine days, we never took a single hot shower?

Two dirty blankets that had been thrown over the slab of cement—our bed—smelled upon our arrival of horse, something that had repulsed us. In truth, several months later, we would have given a lot to recover this healthy odor of a farm, as our blankets had become these cockroach bags that stank of filth and sweat. (Marzouki 2000, p. 68)

Marzouki goes on to recount how prisoners wore the same clothing for three years, ate from broken, dirty dishware, and received the same meals of bread and bad coffee for breakfast, lentils or chick peas or beans and bread for lunch, and vermicelli and bread for dinner. He jokes that when served chick peas, the prisoners played a game of counting the number, from eight for the unlucky to twenty-five for the privileged.

Several decades after Hassan II initiated a process of political reconciliation, his son, Mohammed VI, agreed to establish a Truth and Equity Commission to hear complaints of physical abuse, kidnapping, and detention. The commission received over 22,000 files, either from the victims themselves or representatives. The commission was limited to accounts of abuse during the years 1956 to 1999, provoking criticism that the state was ignoring current abuse, particularly toward Islamists. The director of the commission, Driss Benzekri, a former political prisoner himself, argued that the cutoff in time reflected a change in the regime:

In my opinion, until the end of the 1990s, the then-widespread repressive system had lost its elements and was dissolved voluntarily by the authority itself, which wanted to conform to the new international environment. The late King Hassan II paved the way for these reforms when he initiated a dialogue with the political parties concerning constitutional and political reforms and sharing authority.

In fact, the phenomenon of systematic and repressive disappearance was over by the end of the 1970s. After that, there were separate cases. Repressive detention was used as a kind of direct and deliberate repression against political opposition: Islamists, socialists, Marxists…, etc.

Therefore, the important question for us is, were human rights abuses part of a strategic plan of repression or were they a result of mistakes or breach of law by the security apparatus or by the state itself?

Personally, I opt for the second point because it applies to the status of Morocco. Today, since the end of the 1990s and with the arrival of the new reign, there is a political will, and there are institutions in charge of promoting the respect of human rights. There are laws that have been changed and new ones ratified to cope with the new atmosphere in which Morocco has been involved.[1]

Leaders of Moroccan human rights organizations— including Forum Marocain pour la Verité et l'Equité, the organization Benzekri used to direct fights for recognition and reparations for former political prisoners— criticized the commission for its lack of autonomy from government authority and its inability to lead to prosecutions of those responsible for torture and death. Predictably, Benzekri defended the commission:

This is what a determined group says. It reflects a very narrow ideological point of view. And it is lucky that very few people follow it in our society. The real problem is to improve the situation of the country in a determined historical period, which is the transformation of the political order to the benefit of the country and the society, with the collaboration of both the government and the same society. I think the progress made came mainly from the society, with the support of His Majesty King Mohammed VI and the government. Divergences concerning trials are a purely political debate and have nothing to do with the law. We must take into account what happened in South Africa.[2]

From his perspective, the commission fulfills a purpose, that is to clarify events of the past in order to rectify wrongs and, indirectly, to demonstrate ongoing progress regarding human rights and the institutionalization of political reform.

Regardless of the criticisms of the commission, its existence does indicate a shift in the frontiers of political discourse and action. This chapter discusses these frontiers, describing how they have moved during the period of market liberalization to accommodate new forms of opposition and debate. Whereas during the 1960s and 1970s, the King maintained authoritarian rule, today, a plurality of political actors compete for influence, although always within a kingdom ruled by a King.

During the 1960s and 1970s, King Hassan II faced repeated challenges to his authority and legitimacy from the Socialist and Istiqlal parties and Marxist movements. They criticized his disregard for social progress, his playboy behavior among the jet setters of Europe, his massive wealth, and most of all his view of monarchy as a political rather than a ceremonial institution. King Hassan's insecurity about his opponents led him to establish a regime of little tolerance, one in which opponents were jailed, exiled, or killed.

In contrast, a government of *alternance* came to power toward the end of Hassan II's life, in 1997 (he died two years later). This government consisted of opposition parties led by USFP (l'Union Socialiste des Forces Populaires). As we noted in Chapter 1, the decision to allow the formation of such a government may have represented for the King a way to shift blame onto the opposition for the social problems induced by market liberalization. He also could have acknowledged that only plurality would stabilize a system in a region where radical Islamism had taken hold. Regardless of the motivation, the gradual process of political liberalization initiated by the King has made Morocco distinct within the Arab world.

We could analyze this process of liberalization in Morocco as the consequence of external pressure from international human rights organizations.[3] Certainly, the efforts of European governments and organizations such as Amnesty International to close Tazmamart and eliminate other human rights abuses pushed the late King Hassan II to initiate reforms. Likewise, we could say that the mobilization of domestic actors, for instance, women's organizations, has led the King and successive governments to institutionalize more liberal policies, from reforming the Moudawana, or the family code, to permitting greater freedom of the press. Admittedly, these reforms have succeeded only through the initiative of the King. They do, however, respond to demand within the population.

However, instead of determining the cause of liberalization, we suggest focusing attention on the evolution of political change, where it has been and where it is going. What kind of space for political discussion, participation, and influence exists today in Morocco? More importantly for analyzing the quality of life under globalization, we can ask who participates, who does not, and what the consequences are. Despite genuine progress regarding political institutions and debate, the majority of the population does not engage either directly in elections or indirectly through politicized activity. Younger generations of educated Moroccans express profound alienation from the national mainstream and conventional politics. Moroccans from low-income urban areas have become increasingly drawn to Islamist parties, especially ones such as Al-adl wa al-ihsan that function outside the political system. Moroccans in rural areas turn to community development organizations as well as to traditional political and religious leaders for advocacy.

What then do an attraction to Islamist parties and pervasive alienation tell us about the impact of liberalization on political consciousness? Is the possibility of liberal democracy, so sought after by Western governments such as that of the United States, not convincing to the population? The more immediate response concerns politics itself. First, liberalization in Morocco has remained a carefully managed process that largely implicates elites with a nod toward more transparent elections. Second, and perhaps more important, modern political institutions no longer promise benefits such as social services, jobs, and education that they did in the postcolonial period.

Likewise, the state can no longer rely upon a binding ideology like nationalism for legitimacy. During the oppressive years of the 1960s and 1970s in Morocco, the state signified an object of political desire to its opponents. Marxists and Socialists wanted to reform the state in order to implement their own ideologies of political and social order. Today, as Benzekri has commented, only the Islamists remain idealists. The cynicism and detachment that nonelites convey when referring to the state reflect their lack of identification with the state as an institution that represents their needs and ambitions.

This chapter first analyzes the changes within the political system since independence, ending with an assessment of how the monarchy and the state are transforming within the processes of political and economic liberalization. The objective of existing political parties in the postcolonial era was to determine the nature of the political system, even in opposition to the King. In response, the King became increasingly interested in cultivating a politically loyal bourgeoisie and an economically and socially dependent middle class. He thus defined the stakes for such people, who required patronage to join the bourgeoisie or who needed to accept the existing institutions in order to enjoy social mobility.

Today, participants in the political process want to control the process of market and political liberalization. They want to secure economic interests in an environment in which competition from international firms is increasing, and foreign governments and institutions assert their influence over domestic policy. They also want to influence Morocco's political system, although largely in acceptance of the monarchy as an historic institution. They want to determine if Morocco is going to remain a "strong" monarchy, or if it will become a republic or monarchy governed essentially by a religious guide (similar to Iran [Al-adl wa al-ihsan]), or a Muslim clergy-led party (similar to Turkey [Parti de Justice et Développement]). The other goal, aspired to by leftist and centrist political figures, is a liberal democracy.

In the second section, we outline two other types of political action in Morocco today. The first belongs to new cultural–political movements, which represent an attempt to influence political decisions from outside the political system, or apart from political institutions. These political movements connect international cultural values and transnational political organizations with very local concerns and methods of mobilization.

Unlike the leftist movements of the 1960s and 1970s that concentrated on economic disparity, the new movements, championing particular cultural behaviors and values, have obtained enough popular support to compete with the state for influence. The most notable example of the new movements is the Islamist party Al-adl wa al-ihsan, although women's organizations, human rights organizations, and Berber rights organizations all fit into this category. We discuss these movements in depth in Chapter 3. In this chapter, we juxtapose their existence and power with that of political parties and civil society actors linked to parties and the government.

The third type of political action has been no action at all, at least in the conventional sense. The men and women who fit into this category may sympathize with the Islamists or particular civil society groups, but their political expression primarily assumes forms of alienation and cynicism. They may demonstrate their alienation through the desire to emigrate or through control over body image, and thus individual autonomy. Both trends combine the negative material consequences of market integration with the lure of consumption and global knowledge. They also provoke Islamist political leaders, who view widespread disenchantment with conventional politics as fodder for mobilization and attraction to aesthetics and behavior promoted through the global market as justification for an Islamist regime.

Ultimately, in the chapter, we ask what the process of political liberalization means for politics in Morocco. Who has benefited, who does not participate, and what are the consequences? If the Islamists, regardless of the particular party, gain in popularity, what will this mean for government policy? In addition, what, if anything, will reverse pervasive political disenchantment?

THE PROCESS OF POLITICAL LIBERALIZATION

In the fifty years of independence from the French, the regime in Morocco has changed shape several times, from a contentious modernizing monarchy in the early period to an authoritarian monarchy to an evolving constitutional monarchy in the present day. These changes have reflected the international political climate and the dominant development ideology, as we discussed in Chapter 1. Just as modernization and nationalism buttressed the international and domestic legitimacy of the authoritarian regime of Hassan II, liberalization and cultural and technological globalization have forced the current King, Mohammed VI, and the governing parties to negotiate the scope of their power with political movements and the population as a whole.

Postindependence: 1956–1976

From the first decade after independence (1956), the palace and nationalist political movements competed heavily for control over the political sphere. Mohammed V and the nationalist parties had agreed tacitly to transform Morocco into a modern country with an advanced constitution, but this agreement stumbled over the question of partition of power. For their part, in the years after the declaration of independence, political parties and successive coalition governments quickly demonstrated an inability to implement policy. Between 1958 and 1961, the leftist nationalists and unions governed. These leftists had broken off from the conservative and moderate wings of the Istiqlal party to establish UNFP (Union Nationale des Forces Populaires) and pursue a different policy agenda. The leftists diverged from the rest of the party with respect to nationalization of industry and the public sector, the reclamation of colonized land, nationalization of foreign businesses, and the evacuation of foreign troops. In contrast, the Istiqlal pursued arabicization in education and the judicial system, the reestablishment of the laws of the sharia, and the liberation of all land still under colonial rule. In the end, the leftist government, estranged from the Istiqlal party as well as businessmen and other nationalist figures, faced opposition everywhere.

The battle for power among political parties and between the parties and the monarch ended with the ascent of Hassan II to the throne. The new King applied a vision of authoritarian monarchy that exploited the divisions among his opponents and ended hopes for a modern, democratic political system. The business and traditional elites that had felt alienated from the leftist government decided to support the King. For his part, the King encouraged the creation of new, loyalist parties among Berbers and powerful rural families connected to urban political movements that favor economic liberalism.

The parliamentary system established under Hassan II in 1962 circumscribed further the already constitutionally limited power of deputies. The Chamber of Deputies, though an elected body, had little power over the government or high-level officials. In contrast, the King claimed to be the representative of the people and above criticism or challenge. This system neutralized the left and excluded the UNFP and unions from decision making. The increasing concentration of power catalyzed political reaction during the 1960s (for instance, in 1962 and 1965) that occasionally metamorphosed into active revolt, particularly among youth. The regime not only repressed this opposition but also utilized it to establish authoritarian control for more than a decade, an era referred to as *les années de plomb*, or, literally, the years of lead. In the end, the regime of Hassan II had little in common with the constitutional monarchy promised by his father, Mohammed V, or the elected, egalitarian, theocratic caliph outlined in Sunni Islam.

The personalization of power during this era, which lasted until 1976, was replicated in the public sector and the economy. Rather than push toward a rational, liberal economy, the bourgeois of post-Independence Morocco continued to function through patrimony, receiving and doling out favors, and economic aid based on personal and political connections. The public sector came under this system as well, becoming a source of economic gain for a clique rather than an arrangement serving the public. In all, the economy and public administration resembled the l'Iqtaâ, in which the Makhzen (the traditional regime of the Sultan) distributed resources in order to ensure political loyalty. Political actors thus became economic actors and vice versa, diminishing further the possibility of market efficiency through competition and regulation.

Semiliberalization: 1977–1996

The control of the state over society operated through other channels besides politics, for instance, through religious institutions (the official Conseil de Oulémas) and the surveillance of zaouias, tombs of Muslim saints and spaces for the practice of Sufi Islam (or maraboutisme, so prevalent in Morocco). Through the mid-1970s, those in leadership positions remained intensely loyal to the regime out of necessity. Only sectors in need of technical expertise, such as finance or agriculture, enjoyed a degree of independence from central authority.

Faced with such pervasive networks of loyalty and patronage, by the mid-1970s the opposition had accepted the legitimacy and sacredness of the monarchy. The opposition had also learned to mobilize on the outside. This mobilization not only granted it more popular credibility but also made entering politics formally a risk, as it could lose its reputation of integrity and independence and become like any other disrespected, even despised, political actor. For the most part, despite openings in the political system, the population remained skeptical and thus deliberately apolitical.

The Government of "*Alternance*": 1996–2002

It was not until the early 1990s that the opposition did in fact reach a position of strength. It came to power as the consequence of weak economic growth despite the state's best efforts at market reform. More important, the system of patronage that had sustained the administration and the economy could no longer function as before, simply because there were not enough resources to distribute. Neopatrimonialism was also hindering economic progress. Socially, population growth meant that unrest was inevitable; indeed, Morocco experienced violent demonstrations in the 1980s and early 1990s (notably in 1984 and 1991). Therefore, when Morocco started participating in free trade agreements with Europe, it needed to take additional steps toward institutional democratization in order to head off unavoidable discontent.

This process of democratization could not remain narrow or restrained, especially as social groups affected by liberalization began to place new demands on the state. Unions and opposition parties, soon to come into power, called for democratization on the basis of their popular support. At the same time, the Islamists began to mobilize into a notable and eventually very important political force. The middle class that had benefited from the expansion of education and public sector jobs became more politically critical of the state than the rural elites who had long supported the monarch. These rural elites also lost power in light of the rural–urban migration and institutional restructuring of government, which allowed for regional associations that challenged traditional authority.

When the Koutla, or the opposition government composed of USFP and Istiqlal, came to power in 1997, it did so with the idea that it would ignore political irregularities—for instance, in vote counts—so as not to hinder the transition of power. Despite popular hopes and its long wait for power, the opposition had difficulty surmounting the existing limitations on control, its own divisions, and economic obstacles to demonstrate significant progress.

Mohammed VI

To return to the beginning of Chapter 1, the ascent of Mohammed VI to the throne raised questions about the future of the monarchy. Six years later, with a government of *alternance* still in power, we can ask if the transformation so desired by political, economic, and social actors has actually come to pass. First, how can we evaluate this change—through discourse, through the actors themselves, through events, or through the general atmosphere? Should we question the conflicts of interest, the policies put in place, or the effectiveness of different agencies of government? Second, how do we interpret the transformation of the regime as a whole? Is the political system in Morocco only undergoing superficial change that hides the maintenance of the status quo? Is the monarch in fact consolidating his power? Or is this a transitional period in which traditional interests and practices will continue to lose influence while more modern organizations gain in strength?

One political event that favors the last possibility is the current discussion around revising the constitution. Currently, the constitution does not limit the power of the King, who possesses divine sanction as the Amir al-Mouminine, or commander of the faithful. In Article 19,[4] all division of power in Morocco, even that of the ulemas (the authoritative council within the Islamic clergy), falls below the status of the monarch. The sole social contract represented in the political system in Morocco is that between elected representatives and the population; the King remains above political pacts. Consequently, even if the political sphere has moved toward the institutionalization of a constitutional monarch and the progressive reinforcement of legislative functions, on a practical level, the monarch remains a central actor in the political system.

Until this discussion becomes policy, the architecture of power in Morocco will not change in form. That said, Mohammed VI has continually made overtures toward democratization. He has revisited the various agencies created by his father to give the appearance of liberalization but with the purpose of keeping them from real power; some of the institutions he reinvigorated were the Conseil de la Jeunesse and the

Conseil Consultatif de Droits de l'Homme. The new King also revamped the Conseil Economique et Social (to advise on economic and social issues and national economic strategy) and restructured the Conseil Supérieur and the Conseils Régionaux des Ouléma (national and local councils of religious leaders) in order to ensure the qualification (and political position) of the clergy. In the same vein, the King pushed the government to hasten its efforts to revise laws and the justice system. He also announced revision of the electoral code to promote greater political choice and to guarantee transparency, the reform of the communes (i.e., local governments), and laws regarding public liberties.

These efforts on the part of the King have corresponded with growing dynamism among political elites engaged in civil society, or organizations that lobby for political and social change. Their activism suggests the depth of desire for a break from the patrimonial practices of the past. However, leaders in civil society, unions, and political parties face their own internal conflicts as well as the weight and history of the monarchy. The King, despite his intentions, maintains a parallel government of advisors and friends whose relations extend to those in power in parliament as well as political parties. These intertwined networks and the existence of dual governments incite popular distrust. For all of their noteworthy intentions, civil society and political party leaders have to overcome this distrust to make a significant difference.

Ultimately, to move toward another political system, one in which the distribution of power is clear and the state has the confidence of the people, its political institutions must become more solid and the intermediary bodies, such as the parties, must become stronger. It is difficult to imagine now that the

King will become a monarch alongside and not above parliament. Even if we admit that the King is young and the history of the monarchy long and heavy, and that these factors will affect the pace of reform, the future of the political system in Morocco seems most likely to change owing to pressure from beyond the institutional domain.

CONTESTATION FROM WITHOUT: NEW POLITICAL MOVEMENTS

In contrast to political elites who have become or who have remained attached to conventional political institutions, contemporary cultural–political movements have often acted as the catalysts of political change because of their independent status. Their influence has disrupted the management of liberalization and helped make political change a fragmented, nonlinear process that differs from the careful, moderate advance pursued by most of the political class.

As we discuss in the next chapter, cultural–political movements in Morocco have integrated universalist rhetoric with national and very local popular concerns to mobilize successfully support for their causes, whether they be women's rights, Berber identity, or Islamism. They offer social services, organize marches, publish literature, distribute tapes, and make contacts with similar organizations in other countries. Their power thus lies in their ability to draw upon multiple sources and levels of support, and pushes the far more geographically situated political classes—whose power comes from political appointments, local business, and traditional social hierarchies—into action.

Perhaps no incident illustrates more clearly the influence of contemporary political movements on the substance of political debate and the (mentioned in Chapter 1) management of political liberalization than the indictment in June, 2005, of Nadia Yassine, the daughter of Chiekh Yassine and one of the leaders of his party. The indictment directly referred to an interview given on June 2, 2005, to the news magazine *Al Ousbouia Al Jadida*, but it clearly reflected anxiety over her ongoing comments and her vying for accession to party leadership after her father's death. In the magazine interview, she displayed skepticism about the appropriateness of a monarchy for Morocco. She also predicted that the monarchy would fall in the near future. Several months earlier, she delivered similar remarks, though not as direct, at a conference at UC Berkeley. She belittled political liberalization in Morocco as a reproduction of power that did not address the concerns of the masses:

> The separation of power was practiced by the second caliph after the Prophet. Nothing hinders democracy but the reality—if I speak particularly about the people of Morocco—[is that in Morocco] after independence has been a great theater, a great cinema that we call democratization by the powers in place. We know what Hassenien [after Hassan II] is worth, it is a transplant of democracy that has nothing to do with democracy....[5]

As we cited in Chapter 1, the King reacted to her statements in the interview very quickly, calling her and the two journalists to appear before the courts for antimonarchist remarks.

Her case reflects the political savvy of these new cultural–political movements, their international and transnational character, and their ability to manipulate the substance of political debate.[6] They challenge a political system dominated by overlapping circles of political and business elites through their popular base and through their rhetoric of social and political development. In fact, they often use this rhetoric more effectively than the government or the monarch. In Yassine's case, her statement that she was only offering a "personal opinion" in the interview and her self-presentation as reflective and knowledgeable rather than extremist led both Moroccans and the U.S. government to protest her indictment as an infringement on freedom of speech. The U.S. State Department issued a statement criticizing the regime for disregarding freedom of expression:

> In this case, as with others where the government detracts from freedom of the press or of expression, we are worried. This decision goes against multiple advances accomplished by Morocco in human rights. We consider that freedom of the press is necessary for the consolidation of democracy. In this respect, we encourage the government of Morocco to promote legal reforms that reinforce freedom of expression and the press.[7]

On the other hand, multiple Moroccan political forums and members of the political class, including the PJD, labeled her comments as counter to history and disruptive. The General Secretariat of the PJD issued a statement that read, "Any defamation or denigration of the monarchy regime—that Moroccans chose thousands of years ago—is an unacceptable and irresponsible overstatement."[8] A group of intellectuals, including militants from the 1960s and 1970s, signed a petition entitled "Yes to Democracy, No to Chaos." Abdeljabbar Shimi, in the paper L'Opinion of the party Istiqlal, asked Yassine to what republic she referred: the republic of Egypt under Nasser which killed Sayed Qutb, the spiritual father of contemporary Islamist movements? Or that of Chili under Pinochet, with its "caravan of death" led by the military to silence all opposition? In fact, Shimi insists, the Moroccan people have chosen monarchy because it unites, because it fights against anarchy and feudalism.

In a more formal vein, during a press conference, Mohamed Elyazghi, the premier secrétaire of USFP, described Yassine's commentary as, "categorically unacceptable in the substance and form of the content. It is inconceivable that a responsible citizen can make such declarations, all the more because she made them just after her return from the United States."[9] He later joked that she made her comments after returning from the United States because of the difficulty of obtaining a visa, much less being invited to travel there.[10]

Civil society actors responded as well. Driss Benali, the president of the political forum *Alternative*, denounced her comments as unrepresentative and threatening to genuine reform:

Our goal is to defend a true democratic monarchy that finds itself under several forms of attack. The last to date is that of Nadia Yassine … Not only are her arguments not useful, they also put in danger the progress that has been made in the country and the institution that guarantees national unity … To criticize certain aspects of the monarchy, that is one thing. We ourselves have called for a revision of the constitution that allows for the emergence of a true democratic system. But to be open to the destruction, for the destruction, of a democratic model to which we belong, and we believe in, that is another.[11]

Whereas Benali and others rejected her comments, the politically ostracized cousin of the King, Moulay Hicham, and openly liberal magazines such as *Tel Quel* and *Le Journal* supported her right to free speech. Some in this group also suggested that only a fragile monarchy would feel so threatened by an interview. The editor of *Tel Quel*, Ahmed Benchemsi, mocked the palace's indictment of Yassine.[12] He wrote that "The Moroccan monarchy has set no course. It is therefore at the mercy of the first iceberg that comes along. If an insignificant interview with Nadia Yassine makes it capsize, imagine what a real shock would do."[13]

Although Benchemsi assumes that a confident monarch would ignore such comments, rather than amplify their importance through legal action, he neglects to explain how the globalization of liberal democracy has made such confidence impossible. It is impossible not only because the King may feel threatened by a process of democratization outside his control, but also because the management of such a process requires integrating global, regional, and local forces with the same agility as other, more flexible, more dynamic political actors within Morocco. Yassine can make critical comments concerning the monarchy, garner international support for her right to speak, and place pressure on the King to respond, thus granting her a central position in public space.

The King can only rebuff her remarks, or trivialize their entrance into public discourse, if he has greater capacity to utilize ideas and resources on a global and national level to emphasize the timelessness and relevancy of the monarchy. However, he faces the task of overcoming the weight of his institution as well as mass alienation from conventional political institutions and, thus, significant interest in supporting her indictment. Faced with internatinoal and local political pressures, the King has therefore chosen to focus on specific issues such as the Truth Commission that remain sealed in the past, or women's rights, which appeases international organizations but faces the difficult task of implementation.

ALIENATION IN SPITE OF REFORM

Whereas elites engage in a self-enclosed discourse of political transformation through moderation, and cultural–political movements instigate debate and influence legislation, the majority of the country either shows little interest in politics or declines to participate formally. An analysis of the younger generation's level of interest in politics ironically demonstrates that during the most oppressive years of King Hassan II's reign—the 1960s through the early 1980s—young people confronted security forces by the thousands. Despite an impending election at the time of its publication, the primary conclusion of the analysis was that "Today, the fear (remember when we did not even dare to speak in front of a neighbour for fear that he was a secret agent) has ceded its place to I-couldn't-care-less." Nothing seems to affect this generation: infringements on freedom of the press and therefore on freedom of and thus expression, unemployment, and a lack of available housing. They believe the constitution, parliament, and multiple parties "represent the theatre, a game."[14]

Likewise, in a survey conducted of company executives and managers, the majority, or almost 60%, expressed disillusionment with political parties. Only 3.2% of a sample of 500 belonged to a party. In contrast, a number of those surveyed, or 30.4%, participated in civil society, particularly those living in smaller cities such as Fes or Marrakech.[15] Similar to younger people with far fewer resources, these men (83%) and women (17%) find the political parties stagnant, even archaic. They feel that the central figures remain the same and that parties take up important issues only as the result of external pressures (Islamists, women's associations). They also mistrust elections, despite observers and greater transparency, because of the history of fraud. Again, another survey conducted by Maroc2020 in April 2002 found that 24% claimed to have difficulty distinguishing between parties and 44% stated that they were not at all aware of differences. One-fifth linked themselves to political parties, and 21% believed that political parties reflect fully the needs and priorities of Morocco. Among the respondents, 48% felt that parties share these needs to a limited extent and 37% claimed they do not at all.[16]

Such alienation of different socioeconomic groups, along with the popularity of the Islamist parties, particularly at universities, suggests strongly that political liberalization has not engaged a "public"—that it has not attracted the participation of the people. Why not, if Morocco has advanced political reform further than any other nation in the Arab world? What has the process lacked, beyond factors such as aging leadership and the concentration of power? Indeed, such factors can also inspire mobilization. Instead, younger generations, in particular, have favored actions that incorporate global or transnational aspects such as Islamism and broader trends in migration and communications, to express their desires and frustration.

Successive governments and the King have tried to target alienated younger generations (aged 15 to 34)[17] with economic and social policies and advertising campaigns for job insertion and voting. The government under USFP leadership has also lowered the age of voting to eighteen. The motivation of political parties and the King is clear. Similar to the rest of the Arab world, the majority of the population in Morocco is young. In 2003, 31.9% of the population was under fifteen.[18] Although declining fertility rates mean that the population will gradually become older, the number of young people will put pressure on the job market over the next fifteen years. Over the next decade, 400,000 young men and women will enter the job market each year. In contrast, on average only 200,000 jobs were created each year during the past decade. These jobs also often tended to be part-time and offered minimum or subminimum wages because of increasing competitiveness with other international and domestic firms and the vast pool of available labor.[19]

The pressure on the job market has both political and economic implications. First, frustration and boredom feed unrest and generate support for the Islamists, even if this support means simply that younger generations want to try an alternative form of government, Islamist or otherwise. Second, the gradual aging of the population will require greater funds for social security. Without an expanding tax base from workers, the government will be faced with a shortage of resources to ensure a safety net for older generations.

Policies to decrease unemployment and to promote a sense of social inclusion have included general efforts to improve the educational system, raise literacy rates, and implement electrification and potable water schemes in rural areas. More specific policies have ranged from supporting programs that provide professional training, small-scale funding for entrepreneurs, and microcredit, to funding temporary employment schemes and noneducational activities such as sports. In funding these services, the government has fostered the creation of thousands of associations that work closely with youth.

All of these policies are intended to act as catalysts for greater income generation, social mobility, and self-esteem. They have had mixed results, from the success of girls entering primary school to the failure of the Jeunes Entrepreneurs program, which was supposed to provide capital to young people to help them launch businesses. Moreover, they are policies of self-help intended to facilitate insertion into the job market and combat boredom, but they are not full-scale interventions commensurate with the scale of the problem. They do not resolve the principal issue of improving the quality of education so that government schools would promote and not hinder job prospects. Nor do they acknowledge the number of new jobs needed over the next ten to fifteen years. They also miss the other aspects of education and social stability, namely, transport, housing, and other expenses. According to a recent study, the reasons provided for the rising age of first marriage (27 to 28 years for women and over 30 for men) include a shortage of affordable housing, the lack of available, good-quality jobs, and the desire to marry someone with a stable, well-paying job. The study also mentioned changing values and reform of the family code, the Moudawana, but the respondents deemed economic factors more important than social norms in deciding whether to marry or not.[20]

Despite the stakes involved with government policy and legal recognition of political rights, youth have not generally acted to influence policy direction, except job creation in the public sector. In a memo issued by the USFP after their seventh annual congress in 2005, Mustapha Seyab, a member of the national bureau of the party's youth organization, explains the weak presence of youth in political parties. He blames the history of repression for some of the suspicion of the state and politics. He adds that political parties, despite their proclaimed commitment to democracy, have failed to include youth in decision-making roles. The parties thus echo the dwindling sense of social authority and meaning that market reform has helped accelerate.[21]

The disillusionment with conventional politics has profound implications for both the political system and the parties. Some analysts of disenchantment and anger among younger generations blame the educational system, the policy of the state toward Islam, and the job market for restricting creativity and open-mindedness, and thus fostering intolerance and ignorance. Abdessalam Maghroui refers to Pew Center findings that an overwhelming majority of urban Moroccans between ages 18 and 59 view Christians and Jews negatively and most (60%) believe suicide attacks against the United States are justified. He argues that such opinions reflect the interrelationship between Islam and political power in Morocco and the teaching of fundamentalist, combative versions of Islam.[22]

Other analysts claim that continued concentration of power and the authority of traditional elite families only encourage radicalism and indifference. This concentration of power has resulted in perceptions of distance from decision making and feelings of ignorance about politics. Mounia Bennani-Chraïbi, in her study of politics and identity among young men and women in the early 1990s, interprets this ignorance as a sign of distrust in the country's political legitimacy and a corresponding feeling of isolation, particularly during the end of the oppressive period of King Hassan II's reign.[23] The political environment has evolved since Bennani-Chraïbi conducted her research, but disinterest due to the isolation of political life still exists.

Certainly, the larger environment, inclusive of religious education, the job market, and the privileged elite, has contributed to negative attitudes toward politics. Political institutions have not yet incorporated the masses. The institutions that are directed at the public at large, primarily education and health, have deteriorated under market reform. If we talk of democracy, though, we have to ask what this means, not only in terms of a political structure but also in terms of the identity of the people who would support it. In response, we can go beyond causal analyses of deprivation and alienation to build upon Benanni-Chraïbi's connection between sense of self and political behavior. We can interpret this sense of self not only through negative acts of rejection or disinterest but also through positive creative and social action. For younger generations, this action represents what Susan Ossman calls "lightness." She describes an "en-lightened" body as "one that is without qualities, possessing nothing in particular ... It inhabits no landscape or culture. This body is elusive because it proposes a politics of lack of content based on ideals that do have a history. But this history is a narrative of forgetfulness."[24]

In other words, younger generations do not identify with conventional politics, not only because of endemic problems within the political system but also because they cannot envision themselves within the heavy, weighted national political space. Ossman refers to "heaviness" as "easily held in place."[25] It is a "condition that touches time, mobility, and strength of purpose as much as it reflects a given build [of body]. Heavy, inert bodies are often round, but even thin people can be fixed … [in] timeless spaces."[26] They perceive themselves in this manner because of possibilities that opened up through communications and media and because of the pull of migration in a difficult economic environment. Their manner of expressing themselves thus tends to incorporate both the positive channels of media and the negative juxtaposition of a troubled society and a world full of possibilities.

For example, one of the most popular musicians in Morocco is a young rap star living in Paris who goes by the name Awdellil. Citing him as one of the 50 most influential cultural figures in Morocco, *Tel Quel* praises his lyrics and his anonymity, or his unwillingness to engage in celebrity life. He becomes, in their eyes, an exemplar of "lightness," and thus a figure worth admiring and praising publicly.

All that we know of him can be reduced to a sentence. Noureddine, 22 years, IT student in France. The young man has a certain taste for anonymity and not being able to see his face. With a meager repertoire of three songs disseminated anonymously over the Internet, "Raw Daw," "Messaoud," and "Samia we L'ghalia," he is more famous than any other artist of his generation. And it would not be wrong nor an exaggeration to say that he is also the most talented in his genre. His secret is utilizing unlimited creative capacity, crude language, corrosive humor, and an innate sense of narration. These ingredients made him the idol of millions of young people.

But what above all gave him his aura is the disinterest he attaches to celebrity. "I want to remain free," he is happy to respond to all those who try to resolve the enigma [of his identity]. An enigma that he is ready to keep alive through the next CD that comes out of the shadows.[27]

Awdellil's most popular song is "Raw Daw," which mixes social and political commentary with a survival story about manipulation of social mores:

Don't stop because the words are bad, they show the wisdom of the people. Reflect, meditate … our country can teach a lot, even if it moves between anarchy, corruption, and charlatanism. The people are good but they have a lot of problems.

He got the girl to come to him, he took her for a ride, he nailed her, after long years of frustration. What does she have to complain about? No one forced her to go. What did she believe? That he wanted something else? She didn't have to go.

Do you see now? They made you Raw Daw!

You take it to heart

If you have a pain in your ass, now, you have to just fix it

After all, in this country, everyone screws each other.

The girl goes to see her mother, and she tells her the story. The mother also had a pain in her ass. How is she going to marry off her daughter now? How is she going to get rid of her— and quickly send the younger off? And the father, what is she going to tell him? He is tough, he is going to kill her. Some good souls point her toward a doctor. Some stitches, and the sin is forgotten. The mother pays up everything she has saved—all so that she can finally marry off her daughter.

…

At a party, the girl meets a simple boy, with money, son of a good family. She says to herself: here is the answer! To get it going, she pretends to twist her ankle. The poor boy leaves the party to escort her. Then, she sticks to him like glue … and it ends in marriage. Now, she has the blessing of her mother. Victim, she? Who tricks this poor boy and snares him in her tentacles? And he, is he a man? He passes his time studying, waiting for the girl from a good family. Naïve, pretty, and like him, a virgin.

But, … Raw Daw! He has met the perfect whore. The whole neighborhood has had a turn. If only he had been told. He would not have done it, today, to his grief.

You who listen to me, and who mock me, you make a mistake! You can, you too, be taken. You can have a pain in your ass, too, and you just have to fix it—in silence. Some advice: stop trying to be clever, be modest, don't play it, try to do something good. You believe you are the expert in bad blows, but my poor soul, you have a whole country in front of you! You have to paddle well, the current will always be too strong.

One study conducted in 2004 found younger generations to be "pulled apart," to express the same cynicism and confusion highlighted in Awdellil's music. For instance, one in five respondents had taken drugs while acknowledging that they were forbidden, and the majority of respondents believed in magic and the "evil eye," although not respecting formal Islamic interdictions. Such attitudes could be blamed on the shrinking of available public space, evident at the neighborhood (derb) level, as well as desperation about the job market and, thus, the future.[28]

Perhaps more important, alienation from political life originates in the lack of a sense of existential purpose for younger generations regarding Morocco's future. Creating such a sense of purpose should go beyond policies of cultivating social activities such as sport, or encouraging training and microenterprise. Instead, they should address the "lightness" that characterizes the experience of being young in Morocco, deprived at the local level of both jobs and good schools but engaged in global communications systems and transnational culture. More practically, they should retreat from policies of "incorporation" into an elite-driven political and social system. Rather, they should aim to respond to the desires and identities of the majority.

CONCLUSION

This chapter has discussed three types of responses to the process of political liberalization initiated by Hassan II and pursued further by Mohammed VI. Perhaps the most important move in this process has been the shift of power to the historic opposition, led by USFP and including Istiqlal and PPS. The inclusion of opposition parties in the state has encouraged the evolution of a civil society but has not generated widespread enthusiasm for conventional politics. Filling the void between political leaders and the people, cultural–political movements (from Islamists to women's organizations) have mobilized communities and individuals around particular moral beliefs, legal reforms, and social practices. Their presence has meant that political liberalization has become a process of negotiation, not victory for one party or another, or unilateral imposition of reforms.

After reviewing political change over the past several decades, we can ask if political liberalization has benefited Moroccan citizens and if it has made a qualitative difference in their lives. Certainly, freedom of the press and freedom of expression have expanded. Also, because of the Internet and competition between cultural–political movements, individuals have more choices when it comes to political affiliation.

The King has acknowledged the need to reflect on his position and to encourage the formation of a "citizenry." The question remains if this citizenry will necessarily identify with the state. Will they see the state as representative of their interests? And of what interests in particular—employment, poverty, social security, and education? Or will the loss of connection between nonelite groups and elites undermine any effort to link a public with a governing body? We discuss these issues further in Chapter 4.

Three

In comparing social–political movements of the 1960s and 1970s with those of today, the sociologist Alain Touraine distinguishes between older movements, which fought against social division based on economic distribution, and contemporary social movements that are concerned with the potential of the individual in society. For Touraine, older labor movements reflected underlying structural conflict, between workers and business and political elites, whereas new movements seek control over historical change. He prefers to call new social movements "cultural movements" because they refer to morality and to refinement of the human subject rather than to transformation of a social structure. However, he remarks, these movements still wish to transform social space:

> [I]t is not social transformations or organized social forces that take centre stage, but moral or, as we say more commonly today, ethical demands. Nevertheless, it is moral demands that are in play to the extent that, in legal or other ways, we are actually talking about human rights and the concept of universality of these rights. Is there room for misunderstanding here? A language dominated by interest or strategy has been succeeded by a language dominated by morals, the fear of catastrophe, the often disoriented appeal to something that can resist all violence and cruelty. This is, I think, the essence of the nature of social movements in our society. Should we still talk of social

Cultural–political movements in Morocco do indeed utilize the language of human rights and fulfillment of human capacity. They use the language differently, though. For instance, women's organizations cite verses from the Qur'an in order to build support for incorporating elements of a universalist definition of human rights into law. Conversely, Islamist groups quote verses from the Qur'an in order to deny the need for universalist concepts and terms.

This contrast in logic symptomizes a more fundamental epistemological opposition. New cultural–political movements all focus their attention on the rights of politically marginalized populations, the individual within the social world, and the moral values societies should follow. Yet, their espoused paths to a higher morality diverge dramatically. Whereas the liberal women's organizations want to emancipate the individual from the constraints of society, Islamist organizations want to situate the individual firmly within the collectivity. The fundamental nature of this disagreement and the seeming impossibility of compromise places political parties, and in particular the King, in a tenuous and sometimes unwinnable position.

This chapter discusses, through examples and analysis, how the competition among cultural–political movements in Morocco is framing political discourse and action, and ultimately, contributing to the restructuring of the political system itself. These movements incorporate ideas promulgated at a transnational level into very local action, provoking tensions on a national level that have forced the King and the government into action. However, they are not simply extensions of transnational movements or local movements participating in global politics. They are both more conventional than that and more complex.

In the conventional sense, they are led by middle-class intellectuals who believe their ideological position and their status give them the right to participate in politics. On the other hand, contemporary cultural–political movements differ from older social–political movements, namely, Marxist or labor movements, because the stakes are not necessarily reform of the political system or economic reorganization. Even Yassine's party, despite its demand for an alternative regime, has chosen to remain a critic in opposition with no determined plans for political upheaval.[2] Instead, these movements seek to influence social norms and cultural practices. They concern themselves with the potential of the individual within the social fabric of Morocco.

This chapter shows how such tensions and the fight for control over historical change have led both to institutional and legal reform and to wildly opposing cultural and social trends. More generally, they have made the negotiation of claims for legitimacy and power a centerpiece of conventional politics and cultural debates, linking the two in a manner similar to other countries in the region but in a space that is more open for competition than elsewhere.

The chapter first discusses the origins of political movements and of political action outside of formal politics in Morocco in the postcolonial era. The chapter then describes how two prominent sites of contestation today, women's rights and cultural autonomy for Berbers, demonstrate the power of cultural–political movements over policy. The third section compares the arguments and political tactics of the three most important cultural–political movements: the Islamists, women's rights, and Amazighism. This section also refers to a more minor movement, the movement of Diplomés-chômeurs, which, similar to the others, refers to legal rights, although in a national context and with the specific objective of obtaining employment. The fourth section discusses the other influence on policy, Islamic extremism. Although they constitute only a tiny fraction of Morocco's population, these groups have influenced the passage of legislation and, more fundamentally, the mood of both the public and political elites regarding the present and future of the country's political and social stability.

POLITICAL RADICALIZATION AND MOBILIZATION IN POSTCOLONIAL MOROCCO

The political movements that developed in the colonial and postcolonial eras formed through the prism of the nation-state. They represented, for all of their attachment to the Muslim world or a Marxism that decried the bourgeois manipulation of the modern state, the direct consequence of elite-driven policies of nationalism and modernization. For example, the leaders and members of the two Marxist movements, Ila l-Amam and 23 Mars, were often university students who came from rural areas or low-income neighborhoods in cities. They utilized the press to disseminate their views and, when imprisoned, took advantage of hard-won rights to study and pursue degrees.

Perceiving themselves as agents of history, members of the postcolonial political movements such as 23 Mars believed they had the solutions to the problem of how to emancipate Morocco from the heritage of colonialism and the feudalistic structures that held the country back. They spoke in a modern language of individualism, collectivity, and solidarity when arguing their cause. Similar to the state, they wanted individuals to identify with a collectivity that was both larger than them and their only means of self-fulfillment.

Allal al-Fassi, one of the leaders of the independence movement, stressed the importance of Islam in bringing the new nation-state of Morocco forward. He championed a conservative Islam (*salafiyya*) based on the text rather than the mysticism so prevalent throughout the country. Yet, evoking the possibility of a modern Moroccan nation healed of its social ills, Al-Fassi also subsumed differences between Moroccan ethnic groups such as Arab, Jewish, Berber, and African under a pan-Moroccan identity. He cited Morocco's historic uniqueness as a society relatively uninfluenced by colonialism and imperialism until the nineteenth century as justification for a Moroccan identity not tied to religion or ethnicity. For him, Moroccans retained their integrity as a people for centuries, and with their independence from the French, only had to return to this original "spirit" to achieve progress as a nation.

In keeping with their attachment to the nation-state, members of political movements justified their politics by citing social and economic conditions in a country that had moved well beyond the colonial era. Recalling through poetic letters to his mother his experiences in prison, the leftist militant Salah El Ouadie describes a scene in which he attempts to explain to his torturers his beliefs: "The people live in the darkest misery, I began. Children are not, for the most part, educated; liberties are confiscated; our country, long decades after having achieved independence is still...." His torturers interrupted his explanation, provoking him to consider lying:

> Perhaps this would be more profitable than the truth ... I started to reel off that the people lived in the most comfortable ease, that all children, yes all, went to school, and gave even, for some, classes and conferences, that our adored Morocco itself had been granted by God a liberty that no other country in the world enjoyed, and that moreover, after only thirty small years since independence, it was already...."[3]

Accusing him of not telling the truth, the torturers interrupted him again.

Despite the violence practiced against them, most members of political movements did not support violence against the state but rather strove for political and social means of altering the regime. Al-Fassi, though disappointed in the reign of Hassan II as an authoritarian monarch, did stay on as leader of the Istiqlal party for seven years. Abdellatif Laâbi, who founded the literary and cultural journal *Souffles* in 1966, stated that he never favored the violence practiced by other communist or Marxist groups, commenting that "only the fight of the masses was able to bring about the structural changes (including political ones) that the country needed."[4]

The foundations of political action for Laâbi were the solid social networks created through a common identity and cause. In a paper given to a colloquium on human rights organized by the French prime minister in the mid-1980s, Laâbi demonstrated how the protection of the rights of the individual against the actions of the regime rest with networks of solidarity: "It suffices that it exists so that equilibrium is reestablished between the oppressed and the oppressor, the victim and the jailer."[5] In his personal experience, as a political prisoner during the 1970s, he found original solidarity in the support of the families of prisoners. Mostly illiterate with very little knowledge of politics, women came to the prison (in Kenitra) to provide emotional sustenance for their relatives and later advocated to the world. For Laâbi, those seeking the roots of international solidarity must replicate the "irreplaceable and immediate support that represents the families of the detained. Umbilical cords for the latter, they were equally the best transmission belts between organizations that work for solidarity and victims of repression."[6]

Although leftists like Laâbi have fought for the political mobilization of the masses, they have never gained a large following. They have always represented a small minority that once incurred the repressive fury of King Hassan II. Their movement has never achieved the popularity of Al-adl wa al-ihsan. With political liberalization, former militants have tended to join minority parties on the fringe of politics or create civil society organizations. The Moroccan chapter of Transparency International is but one organization founded by an ex-political prisoner (Sion Assidon). Some individuals, such as Benzekri quoted in Chapter 2, have decided to work for the government.

Unlike their predecessors, contemporary cultural–political movements have reached large sections of the population. They have done so by utilizing multiple methods ranging from social services to tapes and Friday sermons at the mosque. More fundamentally, unlike leftist movements of the twentieth century, contemporary movements have benefited from a particular conjuncture in history, when the state rather than its opponents has to justify its existence in the face of cultural globalization and global market integration. The state has attained legitimacy through binding the identity of the individual with the services and discourse of the state (see Cohen 2004). With this connection undermined by cultural globalization and global market integration (namely, privatization of public services such as education), state actors, specifically the monarch, have had to reflect on their social and political roles. Conversely, cultural–political movements have taken advantage of the impossibility of identification between the singular individual and a geographically and temporally fixed social space. They link the amorphous space of the globe with local experience and thus provide a counter possibility of identity formation. As we explain in more detail in the next section, they work for an order that refers to abstract laws and transhistorical legitimacy.

SITES OF CONTESTATION

Family Law: Islamists versus Liberals

We can interpret the ideological positions and the power of cultural–political movements through two prominent battles: the reform of the Moudawana (the family code) and the struggle for recognition of Amazigh culture and languages. The first fight has pitted the Islamists against liberal organizations, whereas the second has created tension among activists for Amazigh identity, political parties, and the state. Both fights have led the King to act, either by altering the legal code or establishing institutions and allowing them more cultural autonomy. Both fights have also played out through the use of polemical and antagonistic language, as all movements claim to represent the heritage, history, and the future promise of the country.

The Moudawana was originally drawn up after independence in 1956. It represented the only postindependence code in Morocco drawn from Islamic (versus secular) law. The Moudawana was reformed once in 1992 to give women more say over the choice of husband, to dissuade the husband from polygamy, and to provide more assurance of respect, guardianship, and financial support for the wife in case of divorce. However, the code still considered the primary purpose of marriage to be procreation and underscored the authority of the husband. In other words, the wife owed her husband obedience as the head of the household. The code also gave the right to a male relative or the *Wali*, or judge, to speak for a woman without a father and set the minimum age of marriage as eighteen for a man and fifteen for a girl. The husband could still engage in polygamy and repudiate his wife (divorce unilaterally) without the presence of a judge and without notifying his wife.

One women's activist called the 1992 version of the Moudawana a document of bondage:

The Moudawana is really a type of violence, judicial violence, against Moroccan women. It entirely goes against the reality of Moroccan women by not including them as part of society, by not allowing them to participate in development in all areas—political, economic, social, and cultural. When you look at the Moudawana, women are always inferior … The Moudawana is really, in fact, about being your husband's slave."[7]

In less dramatic but no less critical language, Association Marocaine pour Les Droits des Femmes claimed before the 2004 reform that "despite the latest change [1992] that affected a few articles in the Moudawana, the family code continues to regard the woman as an eternal minor."[8] Likewise, the Association Democratique des Femmes du Maroc, in its statement after reform, described the old Moudawana as "based on the submission of the woman in exchange for maintenance provided by the man."[9]

For Al-adl wa al-ihsan and Morocco's Justice and Development Party (the PJD), it was not the family code that necessarily required reform, but rather that Muslims should improve their education about the position of women in Islam. For both parties, the call for change to the Moudawana originated in international politics, specifically the Beijing Conference on Women in 2000, and did not emerge organically from Islam, which has its own tenets regarding the rights of women.

In his writings,[10] Chiekh Yassine argues that women are deprived of the rights originally bestowed upon them, rights that often do not conflict with those of the West but do not mimic them either:

It is urgent to deliver the contemporary Muslim woman, fallen again, perhaps even lower than her pre-Islamic sister, and to draw her up from the abyss of injustice and negligence where she languishes. Our era is perhaps no more merciful toward women than one in which a depraved and inhumane father could cruelly bury his newborn infant if by misfortune it was found to be a girl! ...

Under Islamic law, Muslim women have the right—a right that backward traditions have confiscated from them—to choose their husbands, not to accept a suitor without conditions (including the condition of not marrying a second woman), to ask for divorce, to work and assume social and professional responsibilities, and to dispose freely and independently of their income.[11]

His position and the popular strength of the Islamists, however, did not persuade the governing parties, the women's movement, or, notably, the King.

When King Mohammed VI ascended to the throne in 1999, he made promotion of women's rights a priority. On a personal level, he married Salma Bennani, a 24-year-old engineer from a relatively middle-class family. In the buildup to his wedding, he defied the tradition of sheltering the wife of the King. He presented her publicly and made the engagement and wedding ceremonies media events. Bennani posed for the cover of *Paris Match*, which declared her *enormement belle*.

Politically, in his first *discour royal* as King on August 20, 1999, he declared his intention to improve the position of women in Morocco:

How can we hope to assure progress and prosperity to a society when its women, who make up half of the population, see their rights ridiculed and suffer injustice, violence, and marginalization with disregard to the right to dignity and to equity and yet equity is what our sainted religion confers upon them?[12]

The King subsequently appointed women to high-level political posts and mandated that thirty seats in parliament be reserved for them.

In 2000, after the King announced his program to promote "development for women," women's organizations mobilized to support legal reform, whereas Islamists mobilized against it. Women's associations created an umbrella organization called Collectif Printemps de l'Egalité to lobby the commission appointed by the King to study the reform of the Moudawana. The Collectif also led a march in Rabat in March 2001 near the parliament building that was attended by hundreds of thousands of supporters. At the same time, the Islamists conducted their own march in Casablanca that was attended by far more supporters, perhaps a million.

Both movements led grassroots awareness campaigns. Islamists furthered their strength within university campuses, particularly student unions. Women's organizations in Rabat arranged caravans to go through neighborhoods and communities, rural and urban. They talked to school-age girls and women. One leader of a women's organization said, "We told girls [at a lycée] we wanted to raise the age of marriage to eighteen, and they said, no, twenty-one!" Both sides also put pressure on the government through the media, both local and international. Women's organizations gained support from European and American women's organizations, which engaged in activities ranging from letter-writing campaigns to legal literacy workshops,[13] as well as academic organizations championing the position of women.

In the end, the women's organizations won. First, after the attacks in May 2003, the political climate favored reform over Islamist arguments. Secondly, development organizations such as the World Bank, powerful in its resources and advocacy efforts over government policy, were insisting on progress in women's issues. Lastly, the King himself favored change to the code. The Islamists themselves responded with bland acceptance or muted critique. The head of the parliamentary group of PJD, M. Mustapha R'MID, commented,

> That which characterized the discourse of His Majesty the King was that the King underlined that in his role as commander of the faithful, he could not authorize what God prohibited, nor forbid what His Almighty had authorized. His Majesty the King thus affirmed that these amendments were introduced in taking into consideration the finalities of the Charia that aim to establish justice among people and within the family."[14]

The Islamists also left the debate to the difficulty of implementation. As one journalist put it, "And someone needs to remember to inform the 61% of Moroccan women who are illiterate and the Berber women cloistered in the Rif and Atlas mountains that they have become equal to men."[15]

Legally, the new Moudawana raises the age of marriage from fifteen to eighteen; allows women to claim custody of their children, even after remarriage; discourages polygamy except for exceptional cases; allows spouses to divorce through mutual consent; says that the wife is not subject to the authority of the husband; and, perhaps most important, defines marriage and family as the responsibility of both spouses. The new code also allows the wife, even in the case of remarriage, to sue for custody of her children until a son attains the age of twelve years and a daughter fifteen.

Amazigh Autonomy

The effort by women's rights organizations to change the family code contributes to a larger ambition, driven by international and local NGOs, to create and enforce a system of modern rights. Amazigh organizations and activists take part in this effort, but their strength also reflects the rise of ethnic nationalism at a global level. They draw upon techniques of transnational mobilization, namely conferences, Web sites, and pan-ethnic organizations such as Congrès Mondial Amazigh (World Amazigh Congress), which is based in Paris.

Censuses in Morocco do not record ethnic origin, but Imazighen may constitute half of Morocco's population.[16] The Amazigh population in Morocco stretches over three areas and encompasses three dialects: the Rif in the North, where Tarifit is spoken, the Moyen Atlas mountains, where Tamazight is spoken, and the Haut Atlas and Anti-Atlas mountains along the southern part of Morocco, where Tashlehit is spoken. Since independence, urban migration and arabicization, among other factors, have diluted the strength of language, tribal affiliation, and geography in maintaining Amazigh identity. Younger generations, particularly in urban areas, that regard themselves as Berber in origin may still speak Arabic at home. The state enforces this dilution through laws and cultural circumscription, for instance, the law that prevents families from giving children Amazigh names.

Unlike the effort to change the Moudawana, Amazigh organizations and activists challenge something more abstract and more fundamental, i.e., the Arab-Islamic identity of the regime. For Amazigh organizations, national and global, such an identity limits the scope of self-determination. To these Berber activists, under an Arab-Islamic regime, Berbers remain a minority confined to minor cultural events, without full access to a history that precedes Islam and the Arab empire after Prophet Mohammed. Whatever is the truth of this claim, in light of intermarriage and political glorification of historic Moroccan Berber dynasties such as the Almohades (1130–1269), it symptomizes a grievance against cultural marginalization. This grievance, articulated in the 1991 Charte d'Agadir, calls for revindication through cultural and linguistic rights. Put forth largely by middle-class intellectuals and professionals living in urban areas, the Charte d'Agadir may not necessarily resonate with Amazigh living in rural areas in the Souss.[17] Still, it fundamentally challenges the trajectory of social mobility out of those rural areas (through Arabic and French education) that emerged during the postcolonial era.

Similar to women's organizations, Amazigh activists stage events and utilize the press to champion their cause. For instance, 250 academics, businessmen, artists, and bureaucrats signed a petition in 2000 demanding political recognition and respect for Amazigh languages and culture alongside Arabic and Arab culture:

> As to our Moroccan compatriots who see themselves as proud of being Arab, as we see ourselves as proud of being Amazigh, we believe we constitute with them one sole entity and belong to the same identity. We should not, neither them or us, prevail over the ethnic origins or the lineage of the other, because this justifies mediocrity and the tendency to want to lift up, to enrich, and to impose by subterfuge and not by effort, work, and merit.

Our goal here is none other than to note our will to combat a hegemonic ideology that has fixed as an objective a slow-burning ethnocide that constrains us to see under our eyes a native national language slowly die and a great part of Moroccan cultural heritage erode. Our Maghreb identity cannot be removed from its Amazigh dimension, so profoundly rooted in history, without bringing about irreparable damage. We believe that diversity is a treasure, and difference a factor in the stimulation of human faculties; uniformity, in contrast, only engenders sterility and stagnation....[18]

In direct response to the petition, which was delivered to the *porte-parole* of the King, Mohammed VI established a commission and officially recognized cultural diversity (*Le Discour du Trône* in 2001). The King also created the Institut Royal de la Culture Amazighe (IRCAM) in 2001 and allowed for greater use of Amazigh languages in education and media. The state has encouraged launching cultural events, such as the festival of Amazigh culture in March 2005. From a more socioeconomic perspective, the state has helped cultivate the activities of associations as well as large foundations that work among Berber populations. Medersat. com, the philanthropic program launched by Banque Marocaine de Commerce Exterieur (BMCE) for constructing rural schools, includes instruction in Berber languages.

These successes have corresponded with, and perhaps benefited from, volatile efforts to claim recognition in Algeria. In Algeria, where the Berber population is more concentrated geographically in Kabylie, the mountainous region in the north of the country, success in demanding cultural recognition has extended to forms of political autonomy. Riots in April 2001 in Kabylie led to official recognition of the Berber language. Most recently, in 2005, the Algerian government has decided to dissolve unpopular municipal assemblies, seen by Berbers as extensions of political parties, in favor of a centralized government. These assemblies have also lost power in Kabylie as locally elected councils, known as *Arouch*, have become more powerful.

Although admittedly more fragmented and less politicized than their counterparts in Algeria, Amazigh organizations in Morocco have not obtained the same level of cultural recognition or any political autonomy. Organizations have complained of political indifference toward IRCAM. This indifference keeps IRCAM stunted and weak, without the capacity to channel the aspirations and frustrations of Amazigh activists into broader cultural and ethnic legitimacy. For instance, in addition to the cultural festival mentioned earlier, one of IRCAM's achievements has been the publication of the first comic book in Tamazight entitled *Tagellit N Ayt Ufella*, or the "queen of heights." Unsurprisingly, Amazigh activists in Morocco and Europe regard such actions as insufficient and ineffective. In fact, seven members of the institute resigned in February 2005 in protest over its management.

Without a greater role in politics to moderate their actions, militants have come to demand constitutional revisions that allow for recognition of an Amazigh political party (Parti Démocrate Amazighe), the use of Amazigh languages in education and government administration, the right to secularism, and decentralization of authority, which would allow for more autonomy in Berber areas.[19] The demand for constitutional reform must overcome competition from PJD, among others, who want to emphasize Islam rather than secularism and diversity.

Even if lobbying in the Moroccan parliament does not succeed, the desire for cultural and linguistic recognition, especially among younger generations, and the work of associations in Berber areas have laid the foundation of an irreversible Amazigh movement. These associations, such as Aadec in the central High Atlas or Tiwizi near Agadir, have utilized funds from the government, migrants, and international and national foundations to develop services and infrastructure within their communities. Tiwizi has built roads, installed a generator, improved irrigation, and offered classes in computer training.[20] In this sense, the Amazigh movement is the new cultural–political movement *par excellence*, in that it has connected transnational pan-ethnic politics with the most local and self-determined of actions.

CULTURAL–POLITICAL MOVEMENTS AND THE PUBLIC SPHERE
Common Logic but Different Goals

Although each cultural–political movement champions a separate cause, all challenge the current political and social order. Their fundamental critiques, though centered on different objects, often echo each other in structure. For example, in some cases, they use the common language of rights and moral rectitude. In justifying its claim to righteousness, each movement refers to transnational values, religious and secular. Through this knowledge, they then declare themselves the solution to moral and material problems within Morocco.

For example, Yassine criticizes Western modernity for its materialism, violence, and myopia. He asks a series of questions concerning how Islam can contend with a hegemonic modernity:

What can an idea do? What good is an unarmed idea in the face of a West armed to the teeth?

Aggressive modernity rejects a clear and objective idea when it has no place in the logic of secular discourse, the only sort that is viable. An idea genuinely peaceable, an open and generous proposal of dialogue, is simply *non grata* among the citizens of a culture, of a mentality, of an approachably disdainful economic, political, and military force, fiercely alien to a contrary view that might have the temerity to open blind eyes to the light of day …

What is the point of the barefoot idea, the vagabond proposal, against this colossally concrete, concretely colossal fact?

What is the spiritual and moral value of what an insolvent Islamism has to hawk on the global market over and against the market value guaranteed by the greenback?[21]

At the same time, perhaps the most prominent feminist in Morocco, Fatima Mernissi, describes a world divided by feminine and masculine discourse, a world in which the latter drowns out through its noise any alleged alternative. She suggests that as with Western capitalism for Yassine, masculine discourse allows little space for contradiction simply through the force of its tone:

I prefer to use "loud" over "dominant" because dominant discourse assumes the existence of other discourses that are discordant with or contradictory to it. Now, in Morocco today, in spite of the existence of many masculine discourses (and notably, the discourse of progressive intellectuals who have often broken from the dominant discourse), these have hardly become "listened to." We only hear one sound of the bell, that of loud discourse.[22]

She adds that this discourse "vociferates so much and disposes of such powerful means that it renders inaudible, progressive discourse."[23]

The ensuing strategies for action differ, however, as Yassine answers his own question about how to respond to the authoritarianism of Western capitalism by encouraging a return to the Qur'an, a timeless testament to the authentic, meaningful life. This testament provides "a moment's respite in the midst of modernity's noise."[24] He instructs his readers to "read the Qur'an whenever we can steal a moment's solitude from the promiscuity of modern multimedia. Let us read it whenever informational bottlenecks on the Internet allow us to turn away for a breath of fresh air, to interact with one another as our true selves and not in the guise of modern madness that programs us like robots."[25]

Mernissi likewise criticizes those in power for "sterilizing" alternative, progressive discourse. However, she is concerned with feminist issues, namely polygamy, repudiation (she was writing before the changes to the Moudawana), contraception, and inequality. Progressives are afraid, she writes, of "being accused of atheism by the ideological apparatus dominating all contesting claims."[26] Yassine, who could make an accusation of atheism, takes on contrary issues, for instance—homosexuality and promiscuity. He denounces Darwinism for its promotion of immorality: "As ... evolved apes, homosexuals mass in the streets of Western capitals to claim official recognition of their rights. Homosexual marriage is established in several countries as legitimate and legal. If it is true that the churches are open again in Russia, the avenues of moral degradation are too, alas!"[27]

Through parallel structures of self-description, cultural–political movements ultimately help determine the logic of public debate. Likewise, through the use of similar techniques of recruitment and dissemination, they establish the methods of the contest. Both the logic of debate and shared methods link popular participation with elite leadership and thus forge a political sphere outside of the narrow circles of conventional politics. Yassine tells his readers to "read the Qur'an to put to rest the secularizing insanity which teaches that Islam has nothing to do with politics."[28]

In extending politics outside of the formal political system, contemporary cultural–political movements throw into question the capacity of the monarch and the government to negotiate conflicting ideals and identities. They raise the issue, parallel to mass alienation among younger generations, of identification between the monarch and a parliamentary government on the one hand and a united public on the other. Just as the disenchantment of younger generations reveals the disassociation between state power and notions of personal fulfillment and aspirations, cultural–political movements demonstrate the weaknesses in the authority of the state to make decisions about the cultural and moral fabric of Moroccan society.

Outlining the Substance of Public Debate

The claim to rights—for religion, ethnic autonomy, women, jobs, or otherwise—and the desire to reform public institutions permeates all arguments of cultural–political movements. For example, alienated, unemployed university graduates refer to their right to a job guaranteed by the constitution. Article 13 of the Moroccan constitution reads, "All citizens have equally right to education and to work." Article 12 reads, "All citizens can accede, in the same conditions, to positions within the public sector." Article 13 provides the impetus for a movement calling on the government to resolve high unemployment rates. Article 12 applies particularly to educated, unemployed men and women seeking public sector jobs, which promise security and a stable, albeit relatively low, salary. The Diplômés-Chômeurs repeatedly accuse the government of allowing nepotism and favoritism to influence hiring practices within the public sector rather than meritocracy.

Amazigh and women's organizations also justify the legality of their demands by referring to the constitution, but they go beyond Morocco's borders to the international treaties the Moroccan state has signed, namely, those concerning human rights. They use international organizations and conferences to cite infractions of treaties by the state and to generate foreign support. For example, the text presented by Congrés Mondial Amazigh to the Committee for Human Rights at the United Nations (UN) criticizes the Moroccan government for not respecting the International Covenant on Civil and Political Rights put forth in 1966 and signed by Morocco in 1979.[29] For the Congrés, the Moroccan state has not implemented the right to self-determination (Article 1) and the ban on discrimination (Article 26) laid out in the Covenant on Civil and Political Rights.

In explaining the Moroccan state's failure to comply with these articles, the Congrés cites the findings of the UN Committee on the Elimination of Racism and Discrimination (CERD) in 2003:

In Morocco, Amazigh regions have a standard of living that is largely inferior to that of other regions. Worse, Moroccan authorities have taken it upon themselves to maintain, indeed to accentuate, the economic and social marginalization of these regions, targeting particularly the localities considered the most rebellious. Thus, in the regions of Rif and the Atlas in the southeast and south of Morocco, a great number of independent economic, social, or cultural organizations are voluntarily discouraged by regulatory and administrative obstacles put forth by the Makhzen.

The Congrés also cites infractions of Article 18 of the covenant, which allows for freedom of thought, conscience, and religion:

In the text of the constitution that makes Islam the state religion (Article 6), Moroccans are by authority considered as being Muslim without their having the possibility of choosing, changing, or not having a religion. The "freedom of religion" in the same article does not therefore concern all Moroccans. There is, therefore, no freedom of conscience nor of religion in Morocco and that is contrary to the dispositions of the covenant.

Although not explicit, the text affirms the belief that only a secular regime would allow for adequate self-determination, that a Muslim state cannot fully represent the history of Amazigh culture.

By incorporating the application of international conventions into their arguments for self-determination, Amazigh organizations situate themselves in a position of moral superiority to the state. They play upon this superiority by accentuating victimization and exclusion. For example, the text of the Congrés goes beyond legal claims to a call for protection of Amazigh populations suffering marginalization and despair: "In these territories [Amazigh], all complain of abandonment and desolation. The pauperization of the inhabitants does not cease to advance, notably throwing the younger generations on paths of internal and external exile." The Amazigh want to reverse these trends and regain through self-determination their cultural heritage and their capacity to improve the quality of life in their regions:

> Today and more than ever, the Amazigh want to live with liberty and dignity. Though remaining open to the universal, they wish to be able to enjoy their own identity. Their right to self-determination, this is for them the right to emancipation under a federal, democratic Moroccan state that allows them to assure in freedom their own economic, social, and cultural development.

If the Amazigh assert that the only moral path, one that would allow for the freedom of all of Morocco's inhabitants, is a secular, federal, and democratic state, then the Islamists insist on the opposite, or a state founded on Islamic principles that does not necessarily negate democracy, but does not follow a Western model either.

For Al-adl wa al-ihsan and the PJD, the political systems that leftist and Amazigh movements wish to follow remain in place, in part through a hypocritical repression of religious ideas, or ideas that differ from those represented in political power. In other words, they make a parallel claim to that of Amazigh activists that religious rights are being denied. To realize their aim, the two Islamist parties support an alternative democracy based on the principles of Islam and a citizenship based on Islamic education. In short, they agree with elements of liberal democracy such as a constitution, an independent judiciary, a system of checks and balances between arms of government, and freedom of expression. Yet, these institutions would translate the concept laid out in the Qur'an of *shura*, or deliberative consultation, into modern practice.

Public institutions in Morocco would implement Islamic law, divine law, rather than laws that often reflect the interest of elites over nonelites and the depths of human venality and frailty. Yassine writes, "Democratic forms and methods, applied with precautions and discernment, cannot harm shura; indeed, shura needs them to take effect in the modern world. Only the other face of democracy—the religion of secularism—is unacceptable."[30]

Building Popular Support

The moral contest between ideas and political practices derived from Western modernity and those founded in Islam occurs through very practical intervention into problems of daily life and through abstract promises of existential meaning and individual value. These promises refer to living under the jurisdiction of a state, but they find legitimacy within values that supposedly transcend history, those that have emerged from divine inspiration and the essential qualities of humankind. These twin fronts of cultural–political movements engage individuals before collectivities. They address the contemporary subject who no longer negotiates the boundaries of the nation-state or a traditional culture but, instead, faces the images and ideas that traverse local and global institutions, commodities, and media.

As Susan Ossman (2001) puts it, these are subjects who move through different, sometimes contradictory, worlds. They do not live in one unified world that can be distinguished from another world. They "seek to make existing worlds expand or shrink to gain power or acquire things, but also to develop a sense of how the world in general ought to be."[31] For instance, the young men and women who join ANDCM claim the right to jobs in the public sector based on their education (despite World Bank insistence on cutting these jobs), but they also fight for meritocratic and transparent exam procedures and hiring processes. In other words, they claim privilege and seek equality at the same time. They follow the contradictory path of contemporary social movements outlined by François Dubet, who argues that "the disjunction between actor and system, social integration and the economy, is bringing about … a juxtaposition—rather than combination—of intensified corporatism and far-left radicalism." He explains this juxtaposition, saying that "on the one hand, such movements systematically defend all acquired social benefits and the institutions with which they are linked, often at the cost of maintaining the inequalities they engender. Meanwhile, they have developed just as radical a critique of globalization, conceived as the external cause of all pain and problems."[32]

Alternatively, interweaving modernity with Islam, women's movements situate their battles for modern notions of reform and for greater rights in Morocco within Islamic law. Justifying their practical politics in the transcendence of religion, Islamists support the evolution of certain Western democratic institutions within a Muslim context.

Both move through worlds by building compilations of practice and belief. Women's organizations offer services that protect women before the law and elevate their social and economic independence as individuals. Islamists offer assistance and guidance to socially and economically marginal populations, creating order in places seemingly abandoned or neglected by the state.

In the late 1980s, Aïcha Belarbi, one of the most prominent feminists in Morocco and former ambassador to the European Union from Morocco, decried intellectually substituting labor force participation for genuine equality and social respect:

> Access for women to salaried and productive activity has not engendered profound changes in mentalities. The effects of the economic crisis, [and] the impact of illiteracy have only reinforced social inequality [and] reactivated stereotypes anchored in collective representations of men and women that are expressed in their attitudes, their behavior, and role distribution within the household. To act upon these practices, to transform mentalities and a way of life in the couple and the family, it is necessary to create an economic infrastructure with appropriate competent social services and to give to women and to men the means of realizing their potential and fulfilling themselves in their professional and civic lives.[33]

In the past several decades, among other activities, women's organizations have tried to provide social services that enforce equality before the law and within the family. They provide legal services to women fleeing spousal abuse, counseling to women facing difficulties in their marriages, literacy classes, training classes in handicrafts, and educational seminars on maternal and child health or HIV/AIDS. The objectives of women's organizations borrow from either the discourse of government policy strategies or the directives of development agencies (sometimes their funders) to "improve the position of women" but, behind this discourse, they offer support to their beneficiaries. Their existential promise, therefore, lies not only in greater freedom through income, education, and the protection of the law, but also in greater solidarity among women.

For example, an association in Casablanca called "Centre d'écoute et d'orientation Juridique et Psychologique pour Femmes Agressées" runs a center for legal and emotional support where women can discuss their cases.[34] In the first five years of its existence between 1995 and 2000, the number of visits to the center and the number of cases in which legal assistance was provided both increased almost fivefold, or from 177 to 584 visits and 175 to 610 instances of legal advice. The Association Marocaine pour les Droits des Femmes also manages a center for legal support in Casablanca, created in the 1990s, after the association's members deemed the original reforms to the Moudawana in 1992 unsatisfactory. Similar to the Centre d'écoute, it relies on a network of lawyers to advise and represent women unable to afford legal fees.

The Islamists also translate social critiques into direct social intervention. Unlike women's organizations, they can claim moral advantage because of their distance from the state. They organize their own funding and manage their own resources. They utilize these resources to demonstrate their efficiency and their connection with the people. For example, religious organizations allegedly acted quickly to aid those affected by a fire in Bni M'sik Sidi Othman in 1997 and by a flood in Derb

Sultan in Casablanca in 1996. In a study of Islamic organizations in Morocco, Mohammed Tozy quotes an activist working with the association Assalam (peace). He describes aid given by the association during two catastrophes:

> Thanks to God, we have helped, as an adherent or through the means of our contacts with charitable souls, to relieve a little of the suffering of victims of floods brought on by rain in Derb Sultan in 1996. Thanks to God ... and, how you know it my brother, God says in His book, "Those who spend their money in the path of God resemble a grain that produces seven stalks of which each produces [a] hundred grains." God increases the goods of [those] whom He wants. He is immense and wise. But our most dear wish is to put in contact benefactors with the needy and to no more give strangers [international NGOs] this privilege that the crusaders take advantage of to diffuse their ideology."[35]

For Tozy, this speech revealed not only a desire to do good for other Moroccans, but also to base these acts on religious texts and to compete with international organizations engaged in the same charitable work.

For his part, Yassine writes that the practice of Islam provides the social cohesion missing from Western modernity. "Community direction and morality are at the very basis of the ideal organization of Islamic society, an ideal founded on solidarity and giving ... Returning to the mosque and to the Qur'an will put us back in touch with the ideal of life in community that has been absent from our lives but is ever present and intact in the holy texts and in the hearts of pious men and women faithful to God."[36] He emphasizes that community in Islam forms through service to one another. "In *Islam* [emphasis his], communitarian links are formed through giving and attentive services to one another ... The quality of a civilization is measured by how it treats weaker members: children, the sick, orphans, the poor, the elderly, and oppressed persons."[37]

Unfortunately, the extent of social services provided by religious organizations, whether affiliated with a party or not, remains underresearched in Morocco. According to scholars and journalists, their services largely target university and lycée students, small-business owners and merchants in the informal sector, and families weathering a crisis, similar to the natural disasters mentioned earlier. Al-adl wa al-ihsan has successfully mobilized students in universities over the past decade, in part, through offering parallel education and support services such as photocopying and lending books. The party has also provided capital to small-business owners. Tozy states that hundreds of small-business owners who travel between Nador and Tetuan in the North and Casablanca have benefited from these loans (without interest).[38]

Al-adl wa al-ihsan and the association connection to PJD, Attawhid wal islah (unification and reform), combine their politically directed social actions with religious education. The former possesses a well-structured pyramid organization for imparting the teachings of the sheikh and more closely resembles the historic Sufi brotherhoods of Morocco than the politicized apparatus of the PJD and its associations. Bûsh-ishiya, based near Oujda in the eastern part of Morocco, is the most important religious brotherhood in Morocco. Progenitor of Al-adl wa al-ihsan, the brotherhood claims to have approximately 100,000 members recruited from all groups of society.[39] All these organizations mix religious identity with community support, whether through ritual or social service. They make Islamist movements a vibrant, essential element in the foundation of Moroccan political and social life, a life in which modern, nation-based identity is potentially neither relevant nor attractive to much of the population.

RADICALIZED ISLAM

Unlike cultural–political movements, which influence the political system through techniques ranging from social services to consciousness campaigns, the most extreme of Islamist movements, the Salafiyya Jihâdiyya, focuses its activity on disseminating ideas, creating alternative, combative political communities, and engineering political disruption, sometimes through violence. And unlike cultural–political movements, the movements have increased the authority of the state and the monarch over domestic politics and resulted in increased government surveillance.

For example, the King used the bombings to marginalize the PJD's presence in upcoming local elections. The PJD offered candidates in only 18% of Morocco's 1,544 municipalities. It won 593 seats, or 2.6% of all seats, and did best in Casablanca and Rabat. After the attacks in 2003, the Moroccan parliament passed new antiterrorist laws that authorized, for instance, the right to hold suspects without access to a lawyer and to search homes and businesses without a warrant.[40] The state asserted more control over mosques, particularly those built informally in *bidonvilles* (shanty towns).

Specifically, the state allowed them to open only for prayer (reducing their role as an open public space), forbade opening mosques without permission, and regulated the content of the Friday sermon, so that the imam could refer to the locality but should remain faithful to preset themes. The state also enforced authorization of imams through the Habous (state-run Islamic trust used to fund charitable projects, schools, and so on—much of the trust's resources can come from land) instead of allowing communities to choose their own clerics.

More recently, the King has called upon the Conseil Supérieur des Ouléma to review the issue of fatwas. In his speech to the Conseil, he emphasized that "we will be able to protect and to reaffirm the practice of fatwa … [by] conferring upon it with an institutional character and making in it a collective exercise." In institutionalization, he stresses, "there is neither a place for pseudoclerics in religious matters, or for cults and barefaced charlatans, and still less for the self-proclaimed propagators of hoax allegations."[41]

Ultimately, radical movements are simultaneously more postmodern and more modern than cultural–political movements, particularly Al-adl wa al-ihsan. Whereas Yassine and his followers have developed a movement based on the mystical practice of Islam in Morocco, extremists have borrowed ideas and practices from Saudi Arabia, gained support through the al-Qaeda network, and learned to mobilize in Afghanistan. Whereas many of the followers of Al-adl wa al-ihsan come from middle-income or lower-income families living in working-class or traditional areas (medinas) of large cities, those who participate in radical groups come from the places abandoned by public services, existing outside the institutional realm. Having little positive contact, or even no contact, with the state, these groups reject it, obviously unlike the PJD or even Al-adl wa al-ihsan.

Salafiyya Jihâdiyya, the name for the loose network of clerics and groups practicing a radical, very narrow form of Islam, is dislocated from the mainstream, physically and culturally. Subscribing to the ideology of *takfir*, which calls for rebellion against the corruption of Muslim leaders, this movement shows more interest in the consumption and realization of belief than in gaining influence over specific policies, with the exception of two foreign conflicts: the one between Israel and Palestine and the one in Iraq. Its members choose ideas propagated throughout the world, and they practice them across continents, from Madrid to the United States to Casablanca.

On the other hand, they lack the agility and elusiveness of an organization such as Al-adl wa al-ihsan and, thus, the capacity to manipulate public discourse and the public arena without fear of suppression. Rather, in some ways, they are engaged in a very modern fight, with one side pitted against the other, using violence to repress or destabilize. Although their fight is for liberation, their tactics include self-censorship and intimidation of others.

CONCLUSION

In analyzing the process of political liberalization, we have to ask whether or not a shared framework of action can be translated into a public debate that is able to achieve social consensus, which is viewed by Habermas and others as characteristic of liberal democracy. We can say that although these movements cannot as of yet conduct a constructive dialogue, they also do not represent the dissolution of the public sphere feared by Habermas as a consequence of global media and communications. For Habermas, new media and communications technology may collapse political discussion into individuals and groups seeking reaffirmation of their own political views, rather than engagement with others (see Brothers 2000). Morocco may offer an example of a public arena that places pressure on the state because of a shared language of reclamation and a similar foundation of uncompromising activism that remains largely outside the formal political system. This commonality, as well as the plurality of positions, may represent enough to bind the movements into stable, nonviolent action that forces a political response to social needs and cultural and moral demands.

In addition, contemporary movements have, by forcing the King to make political decisions, also endowed him with a greater role in arbitration over the cultural and political future of the country. Unlike Hassan II during the years of oppression, the monarch is not determining a course of modernization or arabicization that forbids any conflicting or competing ideas. Rather, Mohammed VI and successive governments are making decisions over how to incorporate demands into law and institutions in order both to appease political activists and to maintain the King's own authority. The King and political parties can translate goals of the movements into policy strategy, though in diluted form, and attract leaders of the movements into becoming part of the state. Still, the role of arbitrator and the strategy of co-optation carry risks, namely, a backlash by the losing movement and those still excluded from formal political power. The more politically strategic role for the King—and the position taken by his father, though enacted through repression—is to go beyond the segmentation represented by cultural–political movements and to connect instead social groups that have been increasingly divided through participation in global market integration. We discuss this division in Chapter 4.

INTRODUCTION

In the World Bank's report *Voices of the Poor* (2000), the authors state that their analysis of discussions with the poor, conducted through participatory poverty assessments (PPAs),[1] led to five conclusions. These conclusions ranged from the seemingly obvious—that poverty as a condition extends far beyond a simple economic definition—to the less obvious, namely that nongovernmental organizations (NGOs) have a limited ability to alleviate poverty and face some of the same problems regarding trust, corruption, and accessibility as the state. The last conclusion the authors reached pointed to the continued degradation of the quality of life, particularly because the social networks and social norms that have allowed poor families to survive are deteriorating.

The report highlights the opinions and observations of the people interviewed, and thus does not explore the reasons behind the decline of family systems or values that in another time would have supported those in need. The authors do admit that neither the state nor NGOs—the symbol, in both the developed and developing world, of community action and responsibility—have demonstrated the ability to confront problems such as poverty. Yet, globalization as a development policy depends on the innovation and long reach of NGOs, large and small, as well as the transfer of funds from one section of society, the business class, to another: low-income, socially marginalized groups. How can this happen if the social and institutional infrastructure are lacking? If the state cannot use taxes and nationalistic policies to alleviate poverty and related social conditions, and NGOs can, at best, only provide limited, piecemeal services as a replacement, how will globalization confront inequality beyond the catchall solution of economic growth?

This chapter examines how current development policies in Morocco have, first, altered the social fabric that the World Bank study cites as deteriorating, and second, what this means for the quality of life. Rather than focus solely on the social networks and family systems in low-income communities, the chapter explores how global market integration has affected material and social conditions for the population in general. Furthermore, the chapter discusses how changes in the quality of social and material life have corresponded with the formation of new social groups and reconfigured the relations between these groups during the period of market liberalization. In referring to social life, we not only look at trends such as consumption, we also address identity itself. What kinds of social identities have different groups assumed during the process of market liberalization, and what do these identities imply about the positive or negative impact of development policies?

The first section of the chapter reviews how development organizations and scholars of development have conventionally analyzed the social and material impact of development policies, largely through statistical measurement. The chapter then addresses the issues laid out earlier by offering a more sociological approach to understanding social transformation in Morocco in an era of global market integration. The chapter ends with a discussion of what new patterns of social inequality and social identification mean for the potential of globalization as a development strategy.

MEASURING DEVELOPMENT IMPACT AND DEFINING THE "QUALITY OF LIFE"

Evaluation Logic

International development organizations and scholars of development typically link globalization to economic and social inequality by correlating quantitative data such as poverty figures and economic growth. They also point to examples of political and social action—perhaps local elections or growth in the number of local NGOs—to demonstrate how market integration can provoke other kinds of reform and liberalization and vice versa.

Most international agencies, such as the World Bank or UNDP, rely upon statistical indicators, cross-country comparisons, and descriptions to understand the social impact of reform policies. Human development reports published by the UNDP or the World Development Reports issued by the World Bank track typical measures such as growth in GNP or employment rates as well as more specific measures, such as voting percentages[2] or numbers of PCs.[3]

The World Bank and the IMF analyze trends in statistical indicators against the backdrop of economic liberalization measures. The World Bank Country *Assistance Strategy for Morocco*, written in 2001, makes clear assumptions about the causal logic between growth and poverty and unemployment. The report remarks, "Morocco's growth trend rate [has] continued to slacken, entailing a stagnation in per capita incomes over the last decade, as well as increased poverty and urban unemployment, in sharp contrast to the performance of most other middle-income countries over the same period."[4] The report goes on to comment that the poverty rate at the time, about 20%, and urban unemployment, around 22%, could hinder political reform as well as development efforts. For the World Bank, in other words, economic growth led by liberalization policies is critical to elevating income and reducing unemployment, and thus confronting the challenges of development, including political liberalization.

Evaluations conducted by the IMF and World Bank situate Morocco within a global comparative framework, placing it alongside countries in the same income category in order to measure its progress toward determined objectives that signify "development." Morocco becomes an object of policy and political–behavioral design as well as an indicator of success, slow progress, or failure. For instance, Morocco has long represented a success story for the IMF because of the implementation of its structural adjustment package and the country's aggressive efforts to reform fiscal policy. A summary of the 2001 Article IV consultation concludes the following:

> Executive directors commended the authorities for achieving macroeconomic stability, as evidenced by the comfortable external position and the maintenance of inflation at industrial countries' level, despite adverse weather conditions and terms of trade developments. The economic reforms initiated over the last five years are beginning to yield tangible results as witnessed by increasing productivity, expansion in tourism, and higher foreign direct investment. Directors also welcomed the

authorities' increased emphasis on improving social conditions as well as their efforts to improve governance and transparency in the management of public resources.[5]

In contrast, the World Bank has criticized Morocco for falling behind other countries in the lower-middle income category in the areas of female literacy, maternal health, and rural infrastructure. After completing its country assistance evaluation in 1997, the Bank expressed remorse for not paying more attention to social development. A précis on the report stated, "When compared with countries with similar economic levels, Morocco still rates poorly in terms of primary and secondary school enrollment rates, especially among females. Health indicators are also poor … Morocco's lagging social indicators remain a major obstacle to economic growth."[6] The Bank's tone was softer by 2004, although its dissatisfaction over Morocco's indicators on education, health, poverty, and illiteracy remained intact.[7]

Influenced in part by the work of Amartya Sen, the UNDP takes a more humanistic approach, explaining in the *Arab Human Development Report* (2002) that "income is a means—an important one, but not the only one—in human development … Economic growth is a necessary, but not sufficient, condition for human development. It is the quality of growth, not its quantity alone, which is crucial for human well-being."[8]

Martha Nussbaum, working from Sen's original discussion of expanding human capabilities as an indicator of development, refers to freedoms such as ability to move by choice, to live healthily, and to express political views and organize politically.[9] To her, this approach takes into consideration the quality of life on a universal scale and emphasizes that human beings, when provided with an adequate foundation for individual growth, have the power to reflect and act upon the world. Raising an individual's life to this latter level indicates that processes of development are working, because individuals do not have to worry about survival.

The Depth of Quantitative Measures

In exploring the relationship between economic liberalization policies and confronting poverty and inequality, we have to ask if macrolevel statistical measures or even measures of growth in individual potential are enough to understand the impact of policy measures and what further steps must take place. We could take as an example the World Bank strategy to fight unemployment with economic growth. In the 2004 country brief for Morocco, the Bank states that higher growth rates over the past few years (6.3, 3.2, and 5.2% in 2001, 2002, and 2003, respectively) have corresponded with a slight drop in urban unemployment from 22% in 1999 to 18.3% in 2002 and 19.3% in 2003.[10] In order to lower the unemployment even further, however, the World Bank has stated that growth rates must remain at 6% for years.[11]

These statistics, although revealing of the obstacles facing young men and women entering the workforce, say nothing about the quality of jobs available. Quality implies not only pay and security but also interest and potential. Similarly, higher overall rates of education (57% in 2001–2002) and higher rates of secondary level enrollment (31% in 2001–2002)[12] show that greater numbers of boys and girls are attending school but tell little about the quality of education or the opportunities provided because of that education.

Assessments of development that are statistically based rely upon correlations, or the correspondence between trends. Demographers can point to simultaneous trends in economic growth and declines in fertility and mortality rates. Economists show that upward trends in growth correspond to downward trends in poverty, depending on income distribution levels. In other words, progress in "development" translates into correlation between basic trends—whether in growth, poverty, education, or birth and death rates.[13]

International development organizations as well as scholars draw a global picture of progress between countries by following the same indicators, essentially placing countries on a hierarchical chart of progress toward defined goals. For instance, the World Bank, in its *Country Assistance Strategy for Morocco*, outlines indicators for measuring advancement toward development objectives. The general categories include poverty and inclusion, vulnerability, growth and employment, and public governance. What the bank calls *core performance indicators* represent general goals with targets measured through percentages. Thus, the rural poverty rate should have dropped to 22% by 2004, the maternal mortality rates in rural areas should have fallen to 250 to 270 per 100,000 births, and the overall poverty rate should have decreased to 15%. What the bank calls *monitoring indicators* represent steps toward achieving core objectives. Rural infrastructure programs (water, electricity, and roads) should have reached 60 to 70% of the population and the government should have launched a major literacy campaign in addition to expanding education at the primary level.[14]

The more humanistic measures proposed by Sen, Mahbub ul Haq, the former head of the UNDP, and others still look to universality as the baseline for understanding development.

For the World Bank, countries fit into a category of national income, so that international organizations can make comparisons and state unequivocally that a lower-middle income country such as Morocco should have higher rates of female literacy and female education, or rates closer to those of countries in the same income category. For the UNDP, countries fall within a ranking of human development, measured by the *human development index*. This index measures achievement toward three types of freedoms, freedom being the most basic and most important goal in a humanistic approach. These freedoms are measured through proxy indicators, so that freedom to enjoy a "decent level of living" is measured by real per capita income, freedom to overcome avoidable diseases and premature death is measured by life expectancy at birth, and freedom to have "adequate" knowledge is measured by education indicators.[15] According to *Human Development Report 2004*, Morocco ranked 125th out of 177 countries, with Norway ranking first and Sierra Leone ranking last. For overall human welfare, Morocco ranked below most Arab countries, coming above only Sudan, Djibouti, Mauritania, and Yemen (Iraq was not mentioned).[16]

Yet, what do these measures tell us about transformation in the quality of life in Morocco? We can make comparisons with other countries, have expectations of performance according to tables, charts, and correlations, and point to areas of weakness. What would happen if we went beyond statistical benchmarks of progress and explored first what the statistics mean in terms of areas such as education or health and, secondly, how Moroccans have perceived changes in their welfare both relative to other social groups and relative to the past?

A more sociological approach to analyzing the impact of reform measures would imply understanding the transformation of the life experience of different social groups and the relations between them during the period of market reform. Logically, integrating experiences such as those voiced by the men and women interviewed by World Bank research teams in the PPA exercises with statistical measurements and policy initiatives would touch a wider range of issues than the measurable ones, such as economic productivity or literacy. In other words, analysis that takes into account political, economic, social, and cultural change at several levels, from collective–national to individual, potentially offers a more telling interpretation of the impact of development policy than numerical assessments.

INSTITUTIONAL REFORM AND SOCIAL ACTIVISM

One increasingly prominent method of understanding the impact of economic policies upon poverty and inequality is through analyzing institutional reform and political alliances. This kind of analysis makes the argument that without either infrastructure or allowing for agency from the intended beneficiaries, market reforms cannot succeed as a solution to social problems. Joseph Stiglitz, former chief economist at the World Bank, cites as an example of structural failure the project of an NGO in Morocco. This organization had attempted to train villagers to cultivate chickens, but the project collapsed owing to the absence of a marketing mechanism. The government had faced resistance from the IMF to acting as the distributor and private companies were unwilling to guarantee payment because of the typically high death rates among newborn chicks. For Stiglitz, the market needs institutions that fill the gaps for groups that otherwise would not be able to participate fully, whether through accessing capital, training, or market distribution.[17]

The state's Jeunes Entrepreneurs/Promoteurs program in Morocco, founded to combat unemployment among university graduates, represents an example both of the need for institutions and the existing restraints on institutional effectiveness at regulating the market. Jeunes Entrepreneurs provided 45% of the start-up capital for a new enterprise, whereas national banks, for the most part La Banque Populaire, provided the other 45%. Unfortunately, the government did not provide adequate training or business support for aspiring young businessmen and women to overcome large debts and build a new enterprise. Faced with inexperienced clients and the prospect of default, banks preferred to assist professionals such as dentists or doctors to open offices and entrepreneurs with connections to bank management. Banks regarded loans to professionals as safe, defeating in a way the purpose of providing opportunity to those unable to find capital to take business risks. In addition, with the exception of La Banque Populaire, most banks stopped making the loans out of fear of default. The government then cancelled the program in 2003.

Some students of development and activists in the field have argued that programs such as Jeunes Entrepreneurs fail because they lack public input and the capacity for popular support and participation. These scholars encourage alliances among the state, NGOs, and local populations, with the notion that ideas originating in these communities go further than top-heavy programs. They demonstrate through theoretical arguments and empirical case studies (Rosenfeld and Tardieu

2000; Evans 2002) the benefits of participatory democracy to the process of development. Again, Sen provides a philosophical foundation to the field (see Sen 1999), suggesting that "thick" democracy fostered in public debate and through institutions offers more chances for expanding human potential than institutions restricted in their governance to the interests of political elites. Examples in this field include the effectiveness of participatory city budget management in Porto Alegre (Baiocchi 2001) and the intervention of engineers on behalf of low-income households to secure electricity in France (Ferrari and Tardieu 2000).

In Morocco, examples abound of NGO action to provide basic services or complement those of the state. Fatima Mernissi describes the work of l'Association Migrations Développement in the High Atlas mountains.[18] This NGO was founded in 1986 by North African immigrants returning from France with the expressed goal of developing an infrastructure in their villages. The association ended up providing electricity for forty-five villages. For Mernissi, action on the part of NGOs reflects the same process of participatory democracy championed by American academics seeking examples of power in citizenship as the state becomes more remote. For instance, Jason Ben-Meir, the president of an organization for rural development started by former Peace Corps volunteers in Morocco, calls for the U.S. government to contribute more to community development as a way of building local resources and combating extremism. He claims that $3 billion for community-based projects such as the one facilitated by his organization, the High Atlas Association, would "be sufficient to bring prosperity to Morocco's rural population of 13 million people."[19]

On the other hand, the state in Morocco has also used these associations to implement policy and assert its power, directly and indirectly. The government funds specific programs, such as literacy for adult women, in order to appease institutions such as the World Bank, raise its literacy figures, and to integrate local social activists into the government policy apparatus. Financial insecurity and the absence of medium- or long-term strategies make these organizations particularly vulnerable to government control.

The founder of one of Morocco's oldest community development organizations, now a consultant in Rabat, stated that community development organizations have gone in the past ten years from being charities dedicated to critical and immediate needs to organizations responsible for local social change. She commented, though, that "the state increasingly hides behind them." The former founder and director of an organization called those organizations without their own strategy of development and, thus, a distinctive, well-formed identity, "*femmes de ménages*," or "maids" in French. "I call those organizations '*femmes de menage*,'" he said, in an obviously derisive tone, "because they are not autonomous actors."

What might seem similar to participatory democracy thus needs closer examination to see if the state is responding to activism and becoming more open, or if local organizations are following government needs or both. More importantly, what can we say about the relationship between the political elites and local activists when it comes to popular and state investment in social change for the greater welfare of Morocco? Does local activism represent the newfound involvement and authority of multiple nonelite groups? Or are these groups still subject to the control of the state and ultimately subjugated to the power of political elites?

THE TRANSFORMATION OF SOCIAL SPACE

A response to these questions would center on the relations between the social groups and individuals involved. For instance, teachers on loan from the government typically work as managers in organizations, whether for human rights, development, or otherwise. These former teachers thus work on a local level, directly involved in local issues, whereas members of old families from political and/or business elites form national organizations based in Rabat or Casablanca that explicitly or implicitly place pressure on the state to engage in political liberalization. Do these forums and newspapers represent, though, the maintenance of the hierarchy of power or genuine political change? What does it mean that the local teacher is the local activist, whereas the descendant of a notable family stays in Rabat, the political capital, or Casablanca, the economic capital? What relation do the teachers and the heir have to each other?

Social Transformation within Global Market Integration

Certainly, today, development occurs through the transmission of multiple sets of ideas, some with universalist aims, others derived from local pressures and needs. The impact in a country such as Morocco has been to change the statist orientation of development policy, or nationalist efforts directed at supporting the legitimacy and power of the state, into social fragmentation of both behavior and identities among different social groups. The notion of "belonging" is no longer a powerful glue holding disparate social groups together. And nationalist models of economic and social policy, which favored indigenous firms and expanded state-directed social welfare programs (despite the fact that in practice such policies favored elites), respectively, have withered away since the 1990s.

More profoundly, populations of different socioeconomic backgrounds are no longer connected through the primary myths and policies of nationalism, which are fading or lost. The historian Abdellah Laroui summarizes the power of nationalist logic:

> It seems necessary to distinguish between that which we generally call nationalism and that which is a determined political action by a colonial cadre and the historic matrix that alone gives it [the cadre] its capacity of expansion and penetration. This matrix is the foundation even of local culture; of course, it manifests itself at an intellectual level, but also and above all at the level of behavior; it is a sort of logic that which constitutes society as an organized totality and finality. It is the mirror through which individuals understand their past, their present, and their relations with others ... The veritable cement of society, this matrix is transsocial; it is the common language that all groups speak, even, or rather above all, when they fight each other....[20]

The myths that helped support a nationalist logic after independence included linking fulfillment of individual potential to fulfillment of the nation and establishing an ideal of white-collar employment gained through years of education. King Hassan II claimed in a 1962 *discour du trone* that "our policy aims to emancipate the individual; this is a policy which protects him from ignorance ... It is clear that all development and all progress depend on the extension of culture and the generalization of education."[21]

In order to further this stated goal of "emancipating" the individual, the state implemented five-year plans and policies such as *marocanisation*. This latter policy mandated that Moroccans hold majority ownership in private sector companies. In reality, *marocanisation* benefited elites rather than the middle classes for which it was intended.[22] Social development strategies did provide access to quality education and social mobility for a small portion of the population, but did little to benefit the majority, particularly in rural areas. In 1960, 87% of the total population was illiterate, including 96% of all women.[23] By 1982, 65% of the total population and 78% of all women were illiterate, and 44% of the total urban population and 57% of the urban female population were illiterate.[24]

Today, specific groups have become objects of policies that foster global market competitiveness and reduce state obligation. For example, unemployed university graduates have been told by both King Hassan II and King Mohammed VI to find jobs in the private sector, not in public administration, which the state must reduce in size. The state also must improve literacy, particularly among women, and basic services in rural areas, in part, to improve Morocco's international standing among international development organizations (creditors).

The focus on certain social groups comes from the initiatives sponsored by organizations such as the UNDP. These initiatives to reduce poverty and combat issues such as female illiteracy and weak infrastructure—roads, electricity, and potable water—in rural areas separate out populations facing analogous social and economic problems and then try to rectify them. Arturo Escobar argues that women and small farmers in particular have become subject to the economistic and technological language of development:

> The inclusion of the peasantry was the first instance in which a new client group was created en masse for the apparatus, in which the economizing and technologizing gaze of the apparatus was turned on a new subject. From the late 1970s until today, another client group of even larger proportions has been brought into the space of visibility of development: women … Finally, in the 1980s, the objectifying gaze was turned not to people but to nature—or, rather, the environment—resulting in the by now in/famous discourse of sustainable development.[25]

Escobar refers to vision because, as he argues, the discourse of development "maps people into certain coordinates of control."[26] The purpose of this mapping, according to Escobar, is "not simply to discipline individuals but to transform the conditions under which they live into a productive, normalized social environment: in short, to create modernity."[27]

For organizations such as UNICEF, UNDP, and the World Bank, the programs highlighting issues concerning women and rural communities include digging wells, animal husbandry, microcredit, and maternal care. A UNICEF program in the province of Chaouen, in the northern part of the country, encourages inhabitants to pay a form of insurance to access free essential medication and transportation to a nearby hospital. Another program targets women in rural areas, in which pre- and postdelivery care are provided for pregnant women in order to reduce the risk of maternal mortality. A third UNICEF program installs water sources close to villages to reduce the time that girls and women have to spend carrying it home. A much larger-scale World Bank project invests in the education sector in order to promote basic education for all children through decentralization, improving facilities, expanding teacher support, and setting standards for quality and student performance.

Despite the effort behind these programs, directives such as increasing literacy rates among women or expanding basic education have often occurred without enough thought as to the materials, motivation, and time necessary to teach them effectively, how literacy improves daily life, or how education leads to employment. The former director of an organization that provided literacy courses for women noted, "The state wants women to learn writing, reading, and numbers, in that order. Women want to learn numbers, reading, and writing, in that order." He explained that women wanted to learn numbers for shopping and reading for street signs and everyday practicalities. Writing was far less important.[28] Similarly, the head of a women's organization offering an adult literacy program commented, "Ten months [the government-mandated duration for a literacy course] is not enough to do anything. We therefore have a volunteer who has taken a group of about thirty interested women to continue to study."

The World Bank admits that systemic problems exist in the educational system but does not delve into the reasons why, beyond the need for economic growth to increase job opportunities and the need for infrastructure to support children in remote or marginalized areas to attend school (by building roads, transport, and so on). The Project Information

Document for the Basic Education Reform Support Program lists current "challenges": "Internal inefficiency is high, as evidenced by high dropout and repetition rates. Gender and geographical disparities remain important at all education levels."[29] Without exploring the relation between the job market and education, however, the World Bank misses finding out why dropout rates might be so high, or why families may potentially view girls' education as not beneficial economically or socially. More importantly, The Bank hesitates to acknowledge the scope of reform necessary.

Development organizations do promote "integrated development," meaning approaching development as a combination of improvements in health care and infrastructure, growth in income, and training. Under Wolfensohn, the World Bank introduced the Comprehensive Development Framework, a strategy of going beyond macroeconomic reform and loan conditions to incorporate multiple strategies of attacking poverty, from good governance to microcredit.[30] The National Programme on Sustainable Human Development in Morocco, financed by the UNDP, supports animal husbandry and agriculture, dairy and cheese cooperatives, microcredit, and environmental projects concerning soil and water conservation.[31] The National Programme on Sustainable Human Development and Poverty Eradication in Rural Areas has mapped out regional deprivation through multiple social indicators. Furthermore, the state, with the financial assistance of the World Bank,[32] implemented BAJ1 (*Barnamaj Aoulaouiyat Jtimaiya*) in 1996/1997–2003. The program consisted of interventions in basic health, basic education, small infrastructure, and employment in fourteen provinces regarded as the most deprived.[33] More recently, in mid-2005, the King launched the National Initiative for Human Development and BAJII.

BAJ1 was successful in achieving or surpassing its stated goals. The percentage of girls attending primary school reached 43.9% in 2000–2001, 2.5 percentage points more than the target, the number of health facilities grew much faster than provinces outside the BAJ area, or 57% growth between 1995 and 2002 versus 26%, and investment in roads, potable water, forestation, and so on, created jobs and improved material conditions.[34] However, when asked about their perceptions of the program, the inhabitants of the BAJ provinces responded positively to the interventions but with ambivalence regarding the sustainability of the program. The evaluation report remarks, "The construction of an educational facility without enough teachers, with a limited number of textbooks, furnishings, and canteens, reduces the effect of the project. Likewise, the construction of a sanitary facility without bringing in enough qualified personnel, without medications, ambulance and adequate emergency services, limits the force of impact of the program."[35]

The BAJ and integrated development, for all of their efforts to combine interventions, neglect the larger story of how these interventions relate to the institutions and services available to other populations. For instance, why are doctors and other qualified medical personnel moving or not moving to a rural area? How do the services offered through the program compare to those in urban areas, where many inhabitants of rural areas go for treatment?

We also need to reflect on the motivation and will of beneficiaries, local project managers, and decision-making elites to maintain both specific and multiple interventions. In the discourse and implementation of development policies, low-income communities represent objects of "incorporation" that should become agents within the market-oriented policies of development organizations. In contrast, educated young men and women, who work as volunteers and project managers, receive increasingly little by way of government intervention and support. We can ask what the respective economic and social positions of these populations mean for their relations with each other and, thus, for the local mobilization that Evans and others argue is so critical to sustainable development. We can also question how these groups perceive themselves in relation to the decision-making elites determining government policy and the discourse of progress and the future. Do nonelites feel as though they belong to a national project of social, economic, and political transformation?

Arguably, without situating programs focused on specific populations within the social and institutional contexts, development policies remain temporally and materially limited. They lack the "embeddedness" in social relations that economic sociologists and political economists often talk about when analyzing how markets function (see Granovetter and Swedberg 2001; Evans 1996). In addition, without institutional support for relations between populations separated by economic resources, mobility, and education, social action becomes philanthropic or opportunistic. The process of development thus becomes a function the responsibility and motivation of individual actors and disparate organizations. In other words, it does not become broadly embedded within the larger social structure.

Philanthropic efforts do matter. Othman Benjelloun, the Chairman and CEO of Banque Marocaine du Commerce

Exterieur, founded the BMCE Bank Foundation with a mission to build 1000 schools in rural areas. These schools provide instruction in Amazigh languages as well as Arabic and French and offer after-hours facilities for training and community meetings. Partnering with UNDP, the Ministry of Education, and René Descartes University in Paris, and other foundations in Morocco and Europe, the Foundation has built fifty-five schools in Morocco and contributed to establishing water resources and power supplies in twenty rural areas (*douars*). The scale of this philanthropic effort overshadows other programs, for instance al-Amana, a foundation for microcredit founded in 1997 and directed by Fouad Abdelmoumni, but it belongs to the same category of a powerful businessman responding both to the political desires and needs of the King.[36] In the next section, we explore the subjective and material distance separating this kind of wealthy entrepreneur from a small farmer or an unemployed university graduate that has emerged within market liberalization.

SOCIAL DIFFERENCE AND SOCIAL IDENTITY

In the postcolonial period, social status and economic security could be translated through positioning within the domestic economy—a clerical position in the government bureaucracy versus a trade versus a management position in a national company. Now, we can conceive of the structure of inequality through positioning in relation to the global economy. We can conceptualize the social difference through the economic and cultural resources this positioning provides. Those most implicated in the technological and management aspects of market integration have access to mobility, whether through travel, communications, or the Internet, and to a standard of services compatible with those of other elites in Western or developing countries. Those men and women working in the informal economy or as temporary labor in the government or the private sector have very limited access to mobility and must depend for the most part on public sector services, which suffer from budget cuts, increasing demand, and corruption.

We can explore social transformation within global market integration, and thus the context of development, by distinguishing the political, social, and economic capital different populations maintain, lose, or gain through economic reform. We can interpret the identity of these groups by following the movements of various populations.[37] Tracking these movements implies not only documenting who physically crosses borders but also how people incorporate global images, values, social networks, and technology into their lives. They integrate multiple places, imaginary and real, in an effort to establish their existential space within globalization.

For example, during the 1990s and into the millennium, corner music stores selling copies of CDs and tapes became increasingly popular. The owners of these stores were primarily young men in their twenties to early thirties who doubled as DJs at local clubs and parties during weekends. Their main customers likewise ranged from teenagers to men and women in their thirties who drove by in order to visit and buy CDs. The stores thus became hangouts as well as commercial venues where young men and women could explore the latest Arabic, European, and American pop and dance music.

Instead of analyzing this mélange of music from the perspective of a global recording industry or as a cultural phenomenon encompassing different elements in a global culture, the popularity of mixing music reflects a new, deterritorialized social space. This space has formed because of social alienation and related socioeconomic trends in unemployment, faltering public education, and income disparity. The young men and women who frequent the stores are for the most part high school and university students who have been unemployed or are still unemployed. In the stores, they can go to spaces unattached to the rules of a national culture and social hierarchy or a formal market. Conversely, visiting the stores and acquiring music provides agency that is denied in some form by both the market and the nation.

Music in this case becomes about replacing the existential authority granted in the postcolonial era to these rising generations of the middle class. During the postindependence era, the identity of the middle classes came from participating in modern social and economic institutions such as schools and the government administration. They demonstrated cultural preferences and interpreted their choices through the framework of progress.

In contrast, cultural consumption today reflects the material pressures and the rupture between individual identity and the nation-state that marks our era of globalization. Moreover, these younger generations find existential purpose in the deliberate contravention of national and international institutions and rules.

For one, the vendors, whether working directly on the street or in small stores, typically sell CDs and tapes copied from originals. Therefore, although licensed by the government for sales, their businesses run counter to international law. Secondly, the customers and vendors demonstrate that, despite their marginalization within the Moroccan and global economies, they are the producers of a transnational cultural arena — the consumption of music.

In the next section, we will look at the configuration of social, economic, and cultural resources as well as assets that different groups acquire, retain, or lose during the process of global market integration. We will then see how the new hierarchy of social status and economic power translates into identity within Morocco and the globalizing world.

New Patterns in Social Stratification

The World Bank, IMF, and organizations such as USAID explicitly and implicitly offer ideas about the society they would like to see evolve under market liberalization. This involves, as we suggested earlier, improving the status of impoverished communities through incorporation into the market and modern public services such as education, enlarging a consumer-oriented middle class, and encouraging corporate responsibility toward social issues. In an article encouraging investment in private university education in Peru, the author, an employee at the International Finance Corporation, remarks, "Although telephone lines, cellular services, and software packages help the private sector of developing countries, new educational opportunities—such as UPC—ensure the continued growth of the middle class and lower middle class that is so critical to private sector development, and poverty-alleviating economic growth."[38] Echoing the discourse of the World Bank, *The North Africa Journal*, which reports on business in the region, attributes almost 20% growth in automobile sales in the first quarter of 2005 "to sustained demand and a growing middle class."[39]

What does the middle class mean, though, besides levels of consumption and the required skills to work in the private sector? If we explore the impact of global market integration on social identity and material inequality, we find that the material differences between social groups have grown rather than diminished. More profoundly, the subjective perception of distance may have expanded, as much between social groups as from the binding force of a central project, such as the nation, that had kept them together.

In addition to the obvious factors of income and political power, the gaps now separating elites, the middle class, and lower-income groups are evident in the places they frequent and the institutions they access: private versus public and foreign versus local. And unsurprisingly, the areas they live in are also remarkably different.[40] Ultimately, social groups that form today are based upon participation in the global economy and, inversely, the social status provided by economic position.

The capacity to participate and benefit from global market integration begins with family and education. For example, the Moroccan manager of the year in 2001 for *L'Express*, a French magazine, was Rachid Benyakhlef, the director-general of Sonasid and Managem, two companies in steel and mining that belong to Omnium Nord Afrique, the largest company in Africa and in which the royal family holds a majority share. Benyakhlef attended Lycée Descartes, one of the most prominent schools in the French system, in Rabat, and Louis-le-Grand and the School of Mines in Paris. He inspired admiration from his colleagues and superiors for his ability to work alongside miners. One colleague noted that the evening before the opening of a mine, he was unshaven and wearing heavy miner's boots. "Nothing distinguished him from his men."[41] Benyakhlef claimed that his credo was to "release research from the university ghetto and develop a spirit of innovation."

His goal was to "grow and develop projects. Managing to make a profit doesn't excite me."[42]

Whereas Benyakhlef gains recognition for his technical knowledge and managing skills, others with less valued skills and fewer social and economic resources, such as the DJs, use perhaps similar ingenuity and tenacity to work in the less formal sectors created through global market integration. A Moroccan magazine, *Business*, profiled "cyberpirates" working clandestinely in a large black market in Casablanca, Derb Ghallef. "There are hundreds like Abdessamad [a cyberpirate] in this silicon underground market of Derb Ghallef. Without degrees and from poorer areas of the economic capital [Casablanca], they defy all of the makers of the trendy electronic gadgets quoted on Nasdaq."[43] These young businessmen/hackers/technicians manage to bring electricity into the area without infrastructure. More importantly, they overcome unemployment by creating jobs, however illegal, for themselves. A graduate from an obscure technical institute remarked, "I looked for a job for months without success, and I finally understood that the only solution for me was to launch my own business in the illegal environment of Derb Ghallef."[44] The lack of openings in the formal job market, the absence of recognition for certain degrees, and a paucity of family and other resources leaves this population marginalized socially and economically. Their small, illegal businesses reflect possibility in the new economy, the rigidity of the social hierarchy, and the narrowing passages of social mobility.

These narrowing passages have resulted from trends in unemployment, the concentration of capital, and segmentation in skills and education. Since the advent of market liberalization in 1983, unemployment has risen, particularly among young educated, urban Moroccans, rendering the market for the jobs representing mobility for middle or low-income families far more competitive. The jobs themselves are potentially less secure and lucrative. Unemployment in 2003 among urban youth aged 15 to 24 was 34.5% and among men and women aged 25 to 34, 27.7%. The unemployment rate among those men and women educated to the lycée level and beyond was 27% and among those educated through primary level or in a vocational training program, 26.1%. In contrast, unemployment in urban areas among those without a diploma was 11.3%.[45]

Long-term unemployment has also increased. In 2001, 69% of the unemployed had been jobless for over a year, 48% for three years or more, and 30% for five years or more. In urban areas, the percentage of those unemployed a year or more rose to 76%. Lastly, many of the unemployed, especially among the educated, were first-time job seekers. Four out of five job seekers with a *baccalauréat* or more were looking for a position for the first time, whereas 55.6% of holders of middle-level qualifications and 27.8% of relatively uneducated workers were searching for the first time.[46]

Market liberalization has allowed more young adults to become entrepreneurs and to enjoy the independence and creativity offered by starting and owning a business. The concentration of wealth and intertwining of political and business elites, though, have limited the capacity of small-level entrepreneurs to expand or achieve more influence within the private sector as a whole. These young entrepreneurs typically work in sectors attached to globalization, such as marketing, finance, and advertising. Job growth in other, larger sectors, such as tourism and industry, has been slower. The survey of company executives and managers mentioned in Chapter

2 reported that of the companies involved, most of the new hires in 2000 occurred in commerce (19%), followed by technicians (13%), and accounting (11%). The average number of hires that year among 362 companies was eight.[47]

Those who do find jobs do not necessarily achieve economic security. The average monthly salary of a university-educated government official typically does not allow for savings and investment in children's education, a home, and other goods. Without family support, most bureaucrats or mid-level private sector employees would lack the resources to establish independent, adult lives. This lack of security figures into the system of inequality. The survey cited in Chapter 2 found that gross salaries of company presidents ranged from approximately 350,000 DH a year to 1.5 million DH a year, excluding both social costs such as insurance and perquisites such as transportation. The same survey found that IT officers earned a range of approximately 90,000 DH a year to 630,000 DH a year and an engineer, a range of approximately 90,000 DH a year to 180,000 DH a year.[48]

The capacity for consumption of middle-income and low-income populations has declined despite the expanding availability of consumer goods and the rise of consumer credit companies. Distribution of income has not become more egalitarian since independence, and has worsened since the 1990s. In the 1960s, average annual expenditure per person grew by 4%, but overall expenditure by the wealthiest 10% of the population grew from 8% of total expenditure in 1960 to 29% in 1971. This trend continued during the 1990s, as average individual expenditure declined by 1.9% while the wealthiest 10% continued to consume about 11 to 12% of total expenditures.[49] Expenditure by the poorest 10% of the population remained the same from the 1990s until 2001, or 2.63%, and expenditure by the wealthiest 10% grew from 30.95% to 32.13%.[50]

For bureaucrats or professionals, the cost of living has risen faster than income owing to budget constraints on the public sector. In a dossier on consumption, *News Hebdo* cited a government bureaucrat who complained, "I have worked for thirty-two years in the public sector. In the 1970s, I gave myself permission to buy things for myself and for my family and to travel around to Moroccan cities. Today, this is no longer the case. My salary has only increased dozens of dirhams. We survive like the majority of Moroccans. My three educated children fight to land a job."[51] This bureaucrat has in fact highlighted the differences between generations. Whereas he complains that his secure job no longer pays a wage that enables him to maintain his lifestyle or even keep up with expenses, his children are looking for a job in an extremely competitive market that offers low wages, little security, and minimum benefits, if that.

The occupations of the middle class, with the exception of entrepreneurs or those working for multinationals, have also generally declined in status. For example, teachers express their frustration and disappointment in an educational system burdened by increasing numbers and fewer resources.[52] In a study on public education in Morocco, Assia Msefer describes the feelings of a young teacher on medical leave: "Halim does not believe in his profession. He detests the word *professeur* [because it is] devalued in Moroccan society, he says, by a salary of level nine. He doesn't think to marry because he says his salary is insufficient to take on a house … The absenteeism of students [and] the expression of academic malaise, contributes to the sense of overall abandonment."[53] From her interviews,

Msefer derives three types of reactions to the conditions in the educational system: depression, particularly among younger teachers, defensiveness, and futility.

As social mobility has become less certain, poverty has become more pronounced.[54,55] Although the Haut Commissaire du Plan contends that poverty rates have dropped since the end of the 1990s, or when they reach 19%, to approximately 14% in 2003,[56] these statistics mask the number of people living just above the poverty line who are still in a vulnerable position, socially and economically. In 1999, this figure reached 25% of the population.[57] They also obscure the precariousness of the work that urban poor do find, in the informal sector or otherwise.

At the same time, Morocco has faced criticism from the IMF and the World Bank for not decreasing the minimum wage (daily wage for urban areas and monthly wage for rural areas). Yet, the regulation of these wages is not sufficient to combat poverty. First, many urban workers, and often women, work for less than the minimum wage, because either their labor is not reported to the government or they work longer hours than recorded. Second, the minimum wage applies more to part-time work than full-time work. The CNSS, the agency for social security in Morocco, reports that twice as many part-time employees are paid the SMIG (daily minimum wage) than full-time employees (60% versus 27%).[58] Third, job creation in general has been slow in rural areas plagued by drought and slow economic diversification and in urban areas, where industrialization still remains limited in growth.[59]

As we can see in the difference between someone similar to Benyakhlef and the cyberpirates, social mobility, out of poverty or the lower middle-class, depends upon economic capital and the translation of this capital into educational value. Reproducing their own status, elite and upper-middle-income families often send their children to private elementary schools (to build foreign language skills at a young age) and foreign or private universities. A Moroccan news magazine conducted a small survey of parents who chose to put their children in private schools. The article starts with the comment that "for the parents of students that we were able to survey, public education wallows in structural problems and the educators and teachers in the sector pay less and less attention to the kids."[60] In contrast, private education is supposed to offer higher quality, acquisition of French, and prestige—all assets for the future. Parents place their children in these schools despite the cost, whether 500 DH a month for an average school or 3000 DH a month for one of the international schools. The most sought-after places for children are in the French Mission or the far more expensive American schools, followed by local private schools.

Private education has become so desirable because the public system, which provides education for the overwhelming majority of Moroccans, faces two critical substantive problems in addition to a lack of necessary material resources. First, the system has suffered since the 1970s from political ambivalence about the language of instruction, i.e., French versus Arabic.[61] Bilingualism in French and Arabic separates those able to acquire management positions in the private sector versus those who must either find work in the public sector or face limitations in the private sector. Pierre Vermeren, in his book on education in Morocco and Tunisia (2002), quotes a newspaper interview with a lycée teacher in natural sciences, who remarks sharply, "Do you know that our young graduates coming out of the public school system are generally not capable of aligning two treacherous words in French before forming an unintelligible phrase? Now tell me what business or company worthy of its name would hire this kind of candidate for a job?"[62] Msefer, in her own study (1998), found that many of the students surveyed in her research committed basic errors in French, a problem only shocking because it represents the language of instruction.

Secondly, public education at the lycée and university levels provides vastly inadequate preparation in humanities and social sciences for the current job market. This preparation includes emphasizing rote learning over creativity, so criticized by the World Bank as inadequate for the ingenuity needed to propel growth, and poor teacher training. The most important effect of weak preparation for the job market has been, logically, high rates of unemployment and temporary employment, particularly for those graduates of literature, sociology, and Islamic studies. The second has been increasing interest in private higher education. An unemployed graduate of Islamic studies speaks of his difficulties finding work in *Le Journal d'Emploi*: "I am a young Moroccan, aged thirty-one, with a license in Islamic studies, and I have been unemployed since 1995, and, sincerely, I regret to have chosen this subject; such a subject is worth nothing in the work world." He ends his commentary with advice for future university students: "To conclude, I have a final word to say to young people: the university, for me, is nothing but a factory of unemployment that will never give you a job; this is why I strongly advise you to enroll in practical studies oriented toward work."[63]

Vermeren cites 1984, or the year after the implementation of structural adjustment policies in Morocco, as the true beginning of private higher education. Public universities such as ISCAE, the national business management school, medical, and engineering schools still attract the best students who stay in Morocco. However, French business schools have launched partnership schools, such as Ecole Supérieure de Gestion (ESG) in Casablanca or Sup de Co in Marrakech, that have also attracted students of higher-income families. Like the rest of the educational system, however, those students without financial means, the linguistic skills, or basic education, attend lesser schools that may offer little more than a degree. Perhaps more importantly, many of the better faculties of higher education primarily draw urban students. A survey conducted at the National School of Engineering (EMI) in 1993 found that the number of students born in rural areas had halved since 1981 (from 41.6 to 22%).[64]

In short, higher education for the majority of students only proves beneficial through perseverance in studies and work experience, extraordinary attention to language skills, and family resources. For this majority, higher education, rather than erasing boundaries of stratification, perhaps accentuates them. The students in Msefer's research complained that their course of study contained too much work that was "of little use, boring, not motivating, dated, and far from reality ... Students affirmed that the content of their education was only useful for passing exams...."[65]

In addition to problems in education, basic services in health and infrastructure have not expanded rapidly enough to improve significantly the potential for social mobility and better quality of life. Indirectly, these services do not allow for expansion of individual and family economic resources through decreasing personal costs. Insurance policies cover barely 15% of the population, with three-quarters of those insured working for the state.[66] Government expenditures on health have only increased from 0.9% of GDP to 2% of GDP between 1990 and 2001.[67] And the other side of these statistics is the poor quality of health care available at public hospitals or the meager level of support available for those facing poverty, whether in rural or urban areas.

Stories abound in Morocco of paying for minor equipment at hospitals or "tipping" the staff to obtain better care. Likewise, those in need of care travel to urban areas in order to go to better-equipped hospitals. Ignace Dalle, in his exposé-like treatise of the social, economic, and political problems facing Morocco, quotes cases of men and women seeking emergency services at public hospitals. Describing an incident in which a gardener broke a finger, Dalle writes of the experience, "The waiting room [for emergency services] is dirty and smells bad. Two nurses in a bad mood call the sick in an order that has nothing to do with the chronology [of their arrival]. The least poor succeed in paying a little *bakchiche* [tip] in order to go ahead. No one protests." Of the gardener, Dalle comments, "In his bad luck [to be injured], Bouamar was lucky. He could have lived in a *douar* [rural zone] lost in the High Atlas and would have had to tolerate stoically the pain for days."[68] Recounting another case—a woman suffering from intestinal cancer—Dalle directly cites the opinions of the woman accompanying her: "As had been suggested to me, I had prepared a 'present' for the doctor. To my great surprise, I received a very clear response from him: 'You have the air of being educated and you continue to encourage this type of practice!' The rest of the staff was not so ethical, though. My aunt was hospitalized for six months and it was necessary during this long period to demonstrate generosity toward the personal assistant. If not, she risked being left completely alone."[69]

The same families that send their children to French or American schools in Morocco often rely on private clinics with advanced equipment rather than overburdened, aging public hospitals. The implication for social inequality is again that those families with greater resources have insurance or rely on private health care, even going abroad for treatment. Those without resources depend on a troubled public system. Faced with greater separation between the population benefiting from security and capital and those struggling with public services and limited resources, what can we say about the relationship between global market integration as a strategy of development and fulfilling human capabilities, the approach supported by Sen and the UNDP?

Social Identity within Global Market Integration

Ivan Briscoe quotes a potential migrant from Morocco as saying, "You're considered more illegal in your own country than in any other. You have no work, no health care, no welfare. At least over there you have some protection—all you have to do is get work and you're saved."[70]

The space of separation between elites and nonelites and between young university or professional school graduates and small farmers is framed not only by institutions and material privileges, but also by how these populations identify themselves within globalization. Although Benyakhlef, profiled above, returned to Morocco from France in 1983 because "we could not imagine not returning, to not put ourselves at the service of our country,"[71] the young man quoted above could not wait to leave Morocco for Europe.

Their perspectives reflect the benefits they receive from locality and globalization. Whereas someone similar to Benyakhlef possesses more opportunities in Morocco with his foreign education, the would-be migrant believes he would have a better future with his Moroccan education if he were somewhere else.

Contemporary development policies based on market liberalization have given elites authority within a highly specific location. This authority, ironically, depends upon global resources and corresponds with the diminishing authority of domestically trained middle classes. If we associate this authority with social identity, market reform has acted as a catalyst for the formation of a global middle class marginalized both within an amorphous global arena and a national project (Cohen 2004). The global middle class has emerged as the state has retreated from intervention into social mobility, and thus the support for a middle class. As we can see from new entrepreneurs and the cyberpirates, its rise is also a result of new opportunities produced by economic globalization.

Ultimately, its identity breaks the bond that linked both national and individual emancipation that was so characteristic of the postcolonial period. The inseparability between citizen and state created through modernization policies has fallen apart, leaving in its wake individuals who must struggle through informal and formal means to reach the same modern goal of fulfillment of personal potential. These young men and women seek such personal fulfillment in a world in which institutional and territorial boundaries have frayed, and consequently their goal is less tied to any nationally defined goal.

At the same time, increasing state reliance on local actors, from individuals to associations, in low-income areas to establish and administer social services has separated "community" or "locality" from "nation." We can see the rise of social and political spaces from the Atlas Mountains to the *bidonvilles* in Casablanca that are distinct from the modern nation-state. These spaces form through an intersection of international, national, regional, and local networks and interests (see Daoud 1997). They take shape through the sending and receiving of funds from migrants across Morocco, the Arab world, and Europe, and links to both national and international organizations and institutions. Whereas the state can use local organizations and community leaders to enforce its authority, as stated earlier, these organizations and leaders can also use the state for financial support in order to develop forms of autonomy, from volunteer programs to better locally run basic services. Although this decentralization corresponds with the goals of institutions such as the World Bank, it also challenges the notion of national development and the capacity of the state to represent all communities.

WHAT EFFECT ON DEVELOPMENT?

Located at a global crossroads of ideas, markets, and development plans, Morocco has experienced transformation not only in the organization of its market and social policies but also in the more profound issues of political identity and social structure. Ideas from Bangladesh implemented by U.S.-based organizations such as CRS become programs in Morocco, and although the results in regard to material welfare have varied, the social impact is noticeable. As the result of liberalization policies and related processes of cultural globalization, Morocco has experienced a social transformation that is often unaccounted for in numerically based or universalist evaluations of development. We must consider who now, after several decades of market reform, belongs to the elites, the middle classes, and the poor. In principle, development policies should fortify the middle classes and reduce poverty. Whether or not this result occurs, and it often does not, we still must ask what population constitutes the "middle class" or the "urban poor." How do these groups identify themselves within the process of global market integration and the implementation of development policy?

For example, when lycée students choose to attend private *écoles de commerce* rather than the public university, they do so to learn about the rules of trade and global business and to enhance their chances of finding a job. Without criticizing the growth of private education, we have to ask what the declining prestige and economic value of public education means for opportunity for both the children of middle-income families looking to sustain their status and poorer families wanting to improve their children's welfare. We also have to ask what this means for the legitimacy and authority of the state, the purveyor of such institutions? In this chapter, we have suggested that development policies, whether grand or small in design, do not take sufficient account of the social context in which change is supposed to occur. These policies neglect in theoretical and practical terms—except for philanthropy or for middle-class managers of development and administrative agencies—the ties that bind one social group to another and to the state and, thus, the fabric that supports a broad national project of social, economic, and political transformation.

Summary Remark: What Future for a Development Policy?

In this book, we have tried to show how political and social life in Morocco has changed since the advent of market liberalization in the early 1980s. From the perspective of development policy, Morocco has remained a "good student" of the fundamental principles of neoliberalism. Yet, such subscription to hegemonic development ideology has not translated into adequate progress on social issues. To stave off dissent, the state has initiated a process of political liberalization more profound than any other in the region. The dual processes of liberalization, both economic and political, have generated a series of contradictory positions that now characterize political and social behavior. Elections are increasingly transparent and political choice is growing, as Islamists, Berber activists, and leftists consider participating in the political system. But despite the progress of political reform, the majority of the population remains disenchanted politically. Growing political freedom has corresponded with the rise of radicalism, particularly in impoverished urban areas.

Faced with these contradictions as well as the persistent problems of poverty, low educational levels, gender disparity, and unemployment, we have to ask where development policy goes from here. States in general are now interested in emphasizing social issues. Morocco, for instance, has seen the creation of the National Initiative for Human Development. Development agencies as well no longer call on states to focus primarily on economic reform. These agencies concentrate much of their lobbying, funding, and social programs on improving social indicators and administrative efficiency, particularly in light of the UN Millennium Goals. Will this added emphasis on social progress and governance be enough to decrease poverty and reduce unemployment in Morocco? More importantly, what will these changes, if they happen, mean in terms of quality of life?

For it is not enough simply to raise income or to provide a job. The income must be substantial enough to make a qualitative difference in living conditions and future opportunities. The job must provide not only economic security but also less tangible benefits of self-confidence and self-worth, feelings that counter the desire to participate in extremism. Morocco is far from the only country experiencing the contradictions of liberalization—political liberalization combined with increased radicalism, as well as persistent social and economic problems alongside overly optimistic projections of economic growth. For example, Indonesia over the past decade has made the transition from the long leadership of Suharto to more democratic elections. At the same time, Islamic radicals have gained in strength.

Likewise, an opposition socialist government, similar to the USFP in Morocco, has come to power in Brazil with the expectation of resolving social crises and combating corruption. Yet, two years after his election victory, the government of President Luiz Inacio Lula Da Silva can charm financial circles while fifty-eight million Brazilians still live on less than a dollar per day. Lula can no longer depend on social movements, such as Workers without Land (MST), or unions for political support. Moreover, none of the propositions of the Social Forum held in Porto Alegre have been adopted by the government, from regulation of financial capital to the suspension of debt repayments to protection of the environment. This failure, as well as continued adherence to the liberal market strategies espoused by the IMF, has indicated how difficult change is for any government, regardless of political ideology, that faces entrenched political interests and international economic and political pressure.

Confronted with the political paradox of a socialist government prioritizing economic reform favorable to elites over social concerns, we must question if social concerns can ever become more prominent and less directly tied to economic growth. This question becomes particularly relevant in cases similar to that of Morocco, where rates of economic growth are not sufficient to reduce social problems such as poverty and unemployment. Furthermore, how will development policy foster the connection between the elites that make decisions over wages, public health care, and employment, and the population affected by these decisions? Development policies that target reduction in poverty or a more accountable, efficient administration remain technical. They may claim to encourage public participation in politics through supporting civil society organizations and transparent elections, but they miss the underlying question of how much these organizations or elections actually matter for much of the population.

To answer these questions, development policy must consider the relationship between political institutions and the populations they allegedly represent. It may not be enough to increase transparency. Beyond the occurrence of elections is the larger reason why elections are held in the first place. What can politicians do to improve the lives of their constituents, particularly when the former often live in circumstances entirely different from those most in need of policy intervention? Bill McKibben, an American writer on the environment and development, criticizes Jeffrey Sachs' book, *The End of Poverty*, for ignoring income distribution within nations in discussing strategies that can raise countries' levels of economic growth. McKibben writes that this "orthodox fixation on a country's total GNP makes less and less sense in a world where disparities grow ever sharper. (Take a look at our own country)."[1]

Indeed, McKibben's words may foreshadow a future stage of development thinking: when segregation between countries based on national income gives way to strategizing the reduction of global levels of inequality, both between countries and within them. The unprecedented wealth generated by globalization, as well as the historic level of interaction among countries, economies, and individuals, has not translated into a reduction in inequality. Instead of decreasing, it has only continued to grow. The level of convergence in the quality of life between lower- and middle-income countries and industrialized, high-income countries has remained limited. In 1960, the consumption of the wealthiest 20% of the world's population was 13 times that of the consumption of the poorest 20%; today, this figure has multiplied to 74 times. Whereas 40% of the population in South Asia and 46% of the population in

Sub-Saharan Africa live on less than a dollar a day, only 4% of the population of Norway has a similarly low income.

Globalization has thus resulted not only in economic and cultural integration, but also fragmentation within communities, countries, and regions, a division between those who are "integrated" and the "excluded." The rapid evolution of this inequality has led to the deterioration of the quality of life (not only employment but also means of survival, health, and individual security) and the decline of social cohesion and the culture of the community.

So far, the proposed response to the dangers wrought by this inequality has been insufficient. It is therefore imperative to rethink the conditions under which developing countries open their economies and the role of the international community in this process. This reflection should lead to greater involvement of international institutions in promoting not only the reduction of problems, such as the objective of eliminating extreme poverty laid out in the Millennium Goals, but also formulating a positive conception of quality of life. This conception must include ways of ensuring the financial and political support of wealthier nations, who likewise need to refer to their own issues of poverty and inequality. Lastly, national governments should develop a better method of communicating and asserting their authority with international institutions, particularly concerning the protection of social interests.

At the same time, social groups marginalized within global market integration, from the middle class to low-income urban and rural populations, require more power of expression and activism. For Morocco, a lower-middle-income country with relatively poor social indicators, confronting inequality through international, national, and local interventions is paramount. Returning to the question of political representation, this inequality should be discussed not only as a question of individual or household revenue, but also as the substance of the relations that tie one group to another. The social progress and political overtures pursued in the contemporary development enterprise ultimately depend upon these relations.

Endnotes

PREFACE

1. World Bank Genderstats.
2. Quoted in Laroui 1977, p. 433.

CHAPTER ONE

1. "Interview with King Mohammed VI," *Le Figaro*, September 4, 2001.
2. Discours Royal a l'Occasion du 52ème Anniversaire de la Revolution du Roi et du Peuple, Tetouan, 20 Aôut 2005.
3. Zeghal, Malika, 2005, *Les Islamistes Marocaines*, Paris: Editions La Découverte, p. 7.
4. Ibid.
5. *Washington Post*, Thursday, October 14, 2004.
6. Ibid.
7. Another example could be the state and the King's response to the press when it tests the limits of freedom of expression, particularly on issues such as the war in Western Sahara. The state banned two newspapers founded by Aboubakr Jamaï. Jamaï relaunched the papers under different names but was arrested along with the general editor of *Assahifa al-Ousbouiya* (the new name of the banned paper), Ali Ammar (in 2001), for publishing a story about the finances of Foreign Minister Muhammed Ben Aissa. Jamaï received an International Press Freedom award in 2003.
8. Taken from excerpts from the hearing in "De la parodie á la tragédie," *Tel Quel*, March 7, 2003, p. 18.
9. Ibid, p. 19.

10. "Resist!," *Tel Quel*, June 13, 2003, p. 33.

11. See Rodrik, Dani, 2006, "Goodbye Washington Consensus, Hello Washington Confusion?," paper prepared for the *Journal of Economic Literature* for a critique of 'grand strategies' in development.

12. See Mazur, Robert E., 2004, "Realization or Deprivation of the Right to Development under Globalization? Debt, Structural Adjustment, and Poverty Reduction Programs," *Geoforum*, pp. 61–71.

13. Offenheiser, Raymond C. and Holcombe, Susan H., 2003, "Challenges and Opportunities in Implementing a Rights-Based Approach to Development," *Nonprofit and Voluntary Sector Quarterly*, Vol. 32, No. 2, June 2003, p. 271.

14. Ibid, p. 272.

15. *Declaration on the Right to Development*, 1986, Geneva, Switzerland: The Office of the High Commissioner for Human Rights.

16. Offenheiser and Holcombe 2003, p. 271.

17. Oxfam International Submission to World Bank Review of Conditionality, May 2005; June 14, 2005, p. 3.

18. http://www.worldbank.org/wbi/reducingpoverty/docs/confDocs/JDWShanghaiOpening.pdf, p. 1–2.

19. Ibid.

20. From its Web site, under the section "World Bank/IMF."

21. Brooks, David, "Good News about Poverty," *New York Times*, November 27, 2004, p. 15.

22. From a speech made on July 22, 2004.

23. Maghreb Agence Presse, March 5, 2004.

24. Sachs, Jeffrey, "Don't Know, Should Care," *New York Times*, June 5, 2004.

25. Goldsmith, Edward, "The Last Word: Family, Community, Democracy," in *The Case Against the Global Economy*, San Francisco: Sierra Club Books, 1996, p. 502.

26. From its Web site.

27. From a speech made at the Brazilian Social Forum, Belo Horizonte, Brazil, November 8, 2003. The speech was called "Global Civil Society Instead of Global Civil War."

28. Ibid.

29. The founder of Transparency International (TI) in Morocco is a successful businessman and ex-political prisoner, who fought for the creation of a chapter for years; he supposedly laid the dossier on the desk of the appropriate bureaucrat, who promptly pushed it back to him.

30. The World Bank defines *civil society* as a "wide array of nongovernmental and not-for-profit organizations that have a presence in public life, expressing the interests and values of their members or others, based on ethical, cultural, political, scientific, religious, or philanthropic considerations. Civil society organizations (CSOs) therefore refer to a wide array of organizations: community groups, nongovernmental organizations (NGOs), labor unions, indigenous groups, charitable organizations, faith-based organizations, professional associations, and foundations." Web site information on civil society: web.worldbank.org/WBSITE/EXTERNAL/TOPICS/CSO/0.contentMDK:201014 99~menuPK:244752~pagePK:220503~3piPK:220476~theSitePK:228 717.00.html.

31. United Nations Millennium Declaration, Resolution 55/2, p. 5.

32. Web site information.

33. Cited from *The State of the World's Children*, 2004, UNICEF, p. 4.

34. *Morocco: Livestock and Pasture Development Project in the Eastern Region*, IFAD midterm evaluation report, 1995, p. 6.

35. As of the end of November, 2004, Al Amana had 150,000 active loans for a total amount of more than 364 million MAD.

36. Personal conversation with Shana Cohen, June 2001.

37. The World Bank has provided over $8 billion of loans for 135 projects in Morocco.

38. World Bank 2003, p. 8.

39. Ibid, p. 10.

40. Ibid.

41. See Ketterer, James, 2001, "Networks of Discontent in Northern Morocco: Drugs, Opposition, and Unrest," *MERIP*, pp. 30–33, 45.

42. Morocco has the largest reserves of phosphates in the world, which are used in fertilizers and chemical products.

43. All statistics quoted in this paragraph are taken from the unpublished UNDP report on social development in Morocco.

44. Ministre de Finance and the World Bank.

45. IMF figures.

46. See IMF Article IV Consultation, 2003.

47. *Human Development Report* 2005, Table 14. GDP per capita in Morocco in 2003 was $1,452 and GDP totalled $43.7 billion.

48. Benali et al., forthcoming, "Analyse Sociale du Maroc," UNDP.

49. IMF Article IV Consultation 2005. Table 1, p. 24.

50. World Bank and IMF figures. Public expenditure on education has risen during the past fifteen years, from 5.3% of GDP on education in 1990 to 6.5% in 2000–2002 but expenditure on health has remained the same, i.e., 1.5% of GDP (*Human Development Report 2005*, Table 20).

51. World Bank and IMF statistics.

52. Benali et al., forthcoming.

53. Direction de la Statistique.

54. Van de Walle, Dominique, "Do Basic Services and Poverty Programs Reach Morocco's Poor?," World Bank Discussion Paper, November 2004, p. 16.

55. *North African Journal*, 163, November 22, 2004.

56. http://web.worldbank.org/WBSITE/EXTERNAL/COUNTRIES/ MENAEXT/MOROCCOEXTN/0,,menuPK:294545~pagePK: 141159~ piPK:141110~theSitePK:294540,00.html.

57. See Sweet, Cathy, 2001, "Democratization without Democracy: Political Openings and Closures in Modern Morocco," *Middle East Report* 218, Spring 2001.

58. *Human Development Report 2005* and *Human Development Report Morocco 2003*.

59. Benali et al., forthcoming; and "Morocco's Dismal Prenatal and Maternal Mortality Record," *North Africa Journal*, No. 173, July 21, 2005, p. 12–13; and *Human Development Report 2005*, Table 10.

60. Benali et al., forthcoming.

CHAPTER TWO

1. Quoted in "Driss Benzekri: New Abuses Should Come Within Court Authority," *Morocco Times*, June 1, 2005.

2. Ibid.

3. The oldest human rights organization, Association Marocaine des Droits l'Homme, was founded in 1979 but found it difficult to survive until the political climate became more open in the late 1980s.

4. This article reads, "Le Roi, **Amir Al Mouminine**. Représentant Suprême de la Nation, Symbole de son unité, Garant de la pérennité et de la continuité de l'Etat, veille au respect de l'Islam et de la Constitution. Il est le protecteur des droits et libertés des citoyens, groupes sociaux et collectivités. Il garantit l'indépendance de la Nation et l'intégrité territoriale du Royaume dans ses frontières authentiques."

5. Quoted in *Le Journal-Hebdo*, "Nadia Yassine á Berkeley: un discourse 'trés politique,'" June 20, 2005.

6. Nadia Yassine has a Web site with her comments and the reaction to them (www.nadiayassine.net).

7. Quoted in "Les Etats-Unis Reagissent aux Poursuites contre N. Yassine," *Le Journal-Hebdo*, June 18–24, 2005.

8. Quoted in "PJD Says Islamist Activist Anti-Monarchy Statements 'Unacceptable,'" *Morocco Times*, June 14, 2005.

9. Press conference at 7th USFP Congress, June 2005.

10. Ridouane, Khadija, "Les Paris Gagnés de l'USFP," *Le Matin*, June 24, 2005.

11. From an interview in "Reduire la Fracture Sociale," *Aujourd'hui Le Maroc*, June 13, 2005.

12. Benchemsi and the magazine are, as of this writing, currently in trouble and authorities have fined the magazine almost two million MAD and issued suspended sentences of two months for Benchemsi and Karim Boukhari for multiple counts of slander.

13. Benchemsi, Ahmed, "Le Bal et le Iceberg," *Tel Quel*, June 30, 2005.

14. Trabelsi, Bahaa, "Les Jeunes ne Veulent pas Entendre Parler de Politique," *Citadines*, January 2002, p. 56–57.

15. "Ils Boudent la Politique," *La Vie Economique*, Supplément au numéro 4120, June Friday, 2001, p. 39.

16. Cited in Carnegie Endowment report on Morocco, June/July 2005, p. 12.

17. Benali et al., forth coming, in their report for the UNDP, qualify younger generations as those between the ages of 15 and 34, because this period covers the legal right to work, obtain a job, and marry.

18. *Human Development Report 2005*, Table 5.

19. Benali et al., forthcoming.

20. Ibid.

21. USFP Web site: http://www.usfp.ma/article.php?t=1&id=667.

22. Maghroui, Abdessalam, "La Société se Radicalize dans l'indifference," *Le Journal-Hebdo*, April 2–May 27, 2004, No. 153.

23. Bennani-Chraïbi, 1994, p. 201. She states: "To accentuate knowledge as the foundation of all relation to politics hides not only 'no response' [to a question about interest in politics] but also translates a representation of politics and, above all, a perception of self in relation to this representation: the necessary competence to have one's say on politics does not belong to everyone ... The denial of

interest, the emphasis put on competence, and correspondingly on incompetence, constitutes a form of discrediting even the notion of individual and collective legitimacy to judge and reconsider political authority."

24. Ossman, Susan, 2002, *Three Faces of Beauty*, Durham, NC: Duke University Press, p. 21.
25. Ibid, p. 20.
26. Ibid, p. 19.
27. "Les 50 Marocains les plus Influents," *Tel Quel*, No. 138–139.
28. Houdaïfa, H, "Jeunesse ou le Paradis Perdu," *Le Journal-Hebdo*, June 29–July 1, 2005.

CHAPTER THREE

1. Touraine, Alain, 2004, "On the Frontier of Social Movements," *Current Sociology*, Vol. 52, No. 4, July, 2004, pp. 723-724.
2. The organization is considering forming a party that would participate in elections.
3. El Ouadie, Salah, 2001, *Le Marié*, Rabat: Editions Tarik, pp. 22–23.
4. Laâbi, Abdellatif, 2001, *Les Rêves sont têtus*, Paris: Editions Paris-Méditerranée, p. 12.
5. Ibid, p. 19.
6. Ibid, p. 16.
7. Damia Benkhouya, quoted in "The King and the Chiekh's Daughter," *BBC News*, March 28, 2002.
8. From a brochure explaining the foundation of a legal center.
9. Cited on http://www.mincom.gov.ma/french/generalites/codefamille/Reactions.html, the Moroccan government's Web site on the reform of the Moudawana.
10. We have chosen to cite Yassine's *Islamiser la Modernité* (Al Ofok Impressions, 1998) and its translation, *Winning the World for Islam* (Trans. Martin Jenni, Iowa City, IA: Justice and Spirituality Publishing, 2000—the American publisher for the party). The most prominent and important of Yassine's works, however, is *Al Minhaj Annabaoui*, (The Way of the Prophet). By using the English text, we allow the reader to further explore Yassine's work.
11. Yassine 2000, pp. 93–94.
12. He cited this quote in his *discours royal* in 2001 when he established a commission to explore how to revise the family code.

13. Global Rights, an American organization funded in part by USAID that has a field office in Morocco, offered legal literacy training and human rights education workshops before reform and published a 420-page Arabic-language manual on legal literacy. Western development organizations have also benefited from the reform to the Moudawana. For example, World for International Development was awarded a $770,000 contract from the U.S. State Department to assist in implementation.

14. Reactions to the reform of the Moudawana, Agence Maghreb Arabe Presse.

15. Quoted in Freeman, A. my, 2004, "Re-locating Moroccan Women's Identities in a Transnational World: The 'Woman Question' In Question," *Gender, Place, and Culture*, Vol. 11, No. 1, March 2004, p. 34. Her article discusses the process of reforming the Moudawana and the mobilization of Moroccan women's organizations around universal rights declarations. See Bordat, Stephanie Willman, and Kouzzi, Saida, "The Challenge of Implementing Morocco's New Personal Status Law," Global Rights paper.

16. Maddy-Weitzman, Bruce, 2001, "Contested Identities: Berbers, 'Berberism,' and the State in North Africa," *Journal of North African Studies*, Vol. 6, No. 3, pp. 27–28.

17. See Crawford, David, 2002, "Morocco's Invisible Imazighen," *The Journal of North African Studies*, Vol. 7, No. 1, Spring 2002.

18. *Le Manifeste Berbère*, March 1, 2001.

19. "Que veulent les Berberes?," *Le Journal*, No. 180, October 30–November 4, 2004.

20. This information comes from Shana Cohen's interaction with representatives from Tiwizi. For more, see Roque, Maria-Angels, ed., 2004, *La Société Civile au Maroc*. Paris: Editions Publisud.

21. Yassine 2000, p. xvi.

22. Mernissi, Fatima, 1991, *Le Monde n'est pas un harem*, Paris: Albin-Michel, p. 21.

23. Ibid.

24. Yassine 2000, p. xxv.

25. Ibid.

26. Mernissi 1991, pp. 22–23.

27. Ibid, p. 67.

28. Ibid.

29. Morocco also signed in that year the International Covenant on Economic, Social, and Cultural Rights.
30. Yassine 2000, p. 164.
31. Ossman 2002, p. 153.
32. Dubet, François, 2004, "Between a Defence of Society and a Politics of the Subject," *Current Sociology*, Vol. 52, No. 4, p. 705.
33. Belarbi, Aïcha, 1988, *Le Salaire de Madame*, Casablanca: Editions le fennec, p. 101.
34. As of 2002, Morocco had no shelter for battered women.
35. Tozy, Mohammed, "Les association á vocation religieuse," in *La société civile au maroc*, Paris: Editions Publisud, 2004, pp. 225–226.
36. Yassine 2000, pp. 138–139.
37. Ibid, p. 139.
38. Tozy 2004, p. 237.
39. "Le who's who des islamistes marocains," *Tel Quel*, No. 9, December 24–30, 2001, pp. 16–29, and Tozy 2004, p. 227.
40. See Carnegie Endowment for International Peace, "Morocco," p. 7.
41. First Session of the Conseil Supérieur des Ouléma, Fes, June 7, 2005.

CHAPTER FOUR

1. *Voices of the Poor* defines participatory poverty assessments (PPAs) as an "iterative, participatory research process that seeks to understand poverty from the perspective of a range of stakeholders, and to involve them directly in planning follow-up action. The most important stakeholders involved in the research process are poor men and poor women. PPAs also include decision makers from all levels of government, civil society, and the local elite ... PPAs seek to understand poverty in its local, social, institutional, and political context" (p. 15).
2. UNDP, 2002, *Arab Human Development Report 2002*, New York: UNDP, Table 28, p. 160.
3. Ibid, Table 21, p. 156.
4. World Bank, May 7, 2001, p. 7.
5. Public Information Notice, "IMF Concludes 2001 Article IV Consultation Notice," IMF, June 2, 2001.
6. OED Précis, World Bank, June 1997, pp. 2–3.
7. Country Brief, July 2, 2004.

8. UNDP, 2002, *Arab Human Development Report* 2002, p. 16.

9. Nussbaum, Martha, 1999, "Women and Equality: The Capabilities Approach," *International Labour Review*, Vol. 138, No. 3, p. 235.

10. World Bank Country Brief, July 2004.

11. Ibid.

12. *Human Development Report* 2004, UNDP. The percentage signifies the ratio of enrolled students to that of the total school-age population for that level.

13. For example, World Development Reports demonstrate that economic growth propels poverty reduction by showing tables in which one axis is average annual growth in per capita GDP and the other axis is average annual change in incidence of poverty. See *World Development Report 2000/2001*, p. 48, Figure 3.4.

14. World Bank, Country Assistance Strategy, 2001, R38–39.

15. See UNDP *Arab Human Development Report* 2002, p. 138.

16. *Human Development Report* 2004, Table 1, pp. 139–142.

17. Stiglitz, Joseph, 2002, *Globalization and Its Discontents*, New York: W.W. Norton, pp. 54–55.

18. Mernissi, Fatima, 1997, *Les Aït Débrouille Haut-Atlas*, Casablanca: Editions Le Fennec.

19. Ben-Meir, Jason, 2004, "Create a New Era of Islamic-Western Relations by Supporting Community Development," *International Journal on World Peace*, Vol. 21, No. 1, March 2004, p. 44. He cites the opinions of American and Moroccan NGO personnel and Moroccan officials as corroborating proof.

20. Laroui, Abdellah, 1993, *Esquisses Historiques*, Rabat: Centre Culturel Arabe, p. 131.

21. King Hassan II, March 3, 1962, quoted in Salmi 1985, p. 45.

22. See El Aoufi, Noureddine, 1990, *La Marocanisation*, Casablanca: Les Editions Toubkal.

23. Lahbabi, Mohammed, 1970, *Les Années 80 de Notre Jeunesse*, Casablanca: Editions Maghrebines.

24. *Les Indicateurs Sociaux*, 1993, Rabat: Direction de la Statistique, p. 166.

25. Escobar, Arturo, 1995, *Encountering Development*, Princeton, NJ: Princeton University Press, p. 255.

26. Ibid, p. 156.

27. Ibid.

28. Conversation with Shana Cohen 2002.

29. World Bank 2005, p. 1.

30. For a discussion of the launch of this program under Wolfensohn, see Mallaby and Sebastian, 2004, *The World's Banker*, New York: Penguin Press.

31. UNDP, 2000, *Overcoming Human Poverty: UNDP Poverty Report 2000*, Morocco Country Assessment.

32. The World Bank funded $150 million out of $266 million. The program was launched in 1996/1997 and lasted until 2003. From "Le Premier Rapport d'évaluation de BAJ," *La Vie Economique*, June 6, 2005.

33. See Van de Walle, 2004.

34. "Premier Rapport d'évaluation du BAJ." *La Vie Economique*, June 6, 2005.

35. Ibid.

36. The program is also, at the time of writing, supposedly facing a critical crisis in funding. The cost of maintaining the school buildings and teacher salaries has become too high for the foundation, and thus the entire program is at risk of being shut down.

37. This is similar to what Susan Ossman calls for in her description of the method of linked comparisons. She talks of ethnographic research involving separate places producing knowledge "made by moving through, not over" (Ossman 2001, p. 130).

38. Molavi, Afshin, "Supporting the Private Sector," in "The Knowledge Economy," *Development Outreach*, Washington, D.C.: World Bank, Fall 2001.

39. "Moroccan Auto Market Expands 19.7% Year on Year in 1Q05," *North Africa Journal*, 169th issue, May 3, 2005.

40. One survey of businessmen and professionals found that 65% of ministers, company directors, and political leaders in Casablanca live in Anfa, which is filled with large homes and well-maintained wide avenues (Benhaddou 1997, p. 196).

41. Vidalie, Anne, "Maroc: Le Manager de l'année," *L'Express*, November 15, 2002, p. 54.

42. Ibid, p. 55.

43. Amar, Ali, 2001, "Les cyberpirates de Derb Ghallef," *Business*, No. 2, June 2001, pp. 69–71.

44. Ibid, p. 71.

45. Direction de la Statistique.

46. Bougroum and Ibourk, 2003, p. 347.

47. See Endnote 15 in Chapter 2.

48. "Salaire des Cadres," *La Vie Economique*, June 1, No. 4120, p. 58.

49. Direction de la Statistique.

50. Enquete Nationale sur La Consommation et Les Depenses des Menages 2001/02, Haut Commissariat du Plan.

51. "Le Danger de la Pauvreté s'elargit," *News Hebdo*, No. 4, March 2003, p. 70.

52. Public expenditure on education remained about the same, at around 5% of GDP, during the 1990s. At the same time, net enrolement, meaning the number of children enrolled as compared to the number of children of the official school age for that level, rose for primary education from 57% in 1990/1991 to 88% in 2001/2002. Spending on primary education did rise from 34.8 to 48% from 1990 to 1999–2001 of total spending [spending on tertiary education dropped from around 16% to almost nothing]. From *Human Development Report* 2004, Table 10.

53. Msefer, Assia Akesbi, 1998, *Ecole, Sujets et Citoyens*, Casablanca: Edit Consulting, pp. 223–224.

54. The Haut Commissaire du Plan has estimated that the poverty rate declined between 1999 and 2003 to 13.8% because of economic growth, particularly in agriculture (reducing rural poverty). However, as Benali et al. (forthcoming) suggest, this rate reflects an adjustment in the definition of poverty (income in MAD).

55. Masmoudi, Khadija, "Pourquoi la pauvreté a explosé depuis 1993," *L'Economiste*, November 2001, p. 10, and Benali et al., "Analyse Sociale du Maroc," forthcoming report for UNDP.

56. World Bank statistics are 15% in 2004.

57. Benali et al., forthcoming.

58. Ibid.

59. The percentage of manufacturing in GDP declined from 18.5% in 1984 to 16.5% in 2004 (World Bank statistics).

60. *Finances News Hebdo*, Hors Serie No 4, March 2003, p. 65.

61. Responding to the 1995 World Bank report, which criticized the government for not planning education to coordinate with the needs of economic growth (a skilled labor force), Hassan II set up in 1996, l'Option Langue Française in several subject areas at secondary school level with the idea of reintroducing French as a language of instruction.

62. Quoted in Vermeren 2002, pp. 460–461.

63. "La galére d'un licencié au quotidian," No. 6, January 15–21, 2001, p. 20.

64. Cited in Vermeren 2002, p. 481.

65. Msefer 1998, p. 152.

66. Jaïdi, Larbi, paper entitled "Défis et conditions du développement social au Maroc," p. 9.

67. *Human Development Report* 2004, Table 19.

68. Dalle, Ignace, 2001, *Maroc 1961-1999: L'espérance brisée*, Paris: Maisonneuve et Larose, p. 105.

69. Ibid, p. 107.

70. Briscoe, Ivan, "Dreaming of Spain: Migration and Morocco," *Open Democracy* Web site, May 27, 2004.

71. Vidalie, p. 54.

CONCLUSION

1. McKibben 2005, p. 24.

Bibliography

Aujourd'hui le Maroc.

Business.

Citadines.

Finance News Hebdo.

L'Economiste.

L'Express.

La Vie Economique.

Le Figaro.

Le Journal Hebdomaire.

Le Matin.

Morocco Times.

New York Times.

North African Journal.

Tel Quel.

Declaration on the Right to Development. 1986. Geneva, Switzerland: The Office of the High Commissioner for Human Rights.

Baiocchi, Gianpaulo. 2001. "Participation, activism, and politics: The Porto Alegre case and deliberative democratic theory." *Politics & Society,* Vol. 29, No. 1, pp. 43-72.

Bates, Robert H. et al. 1998. *Analytic Narratives.* Princeton, NJ: Princeton University Press.

Belarbi, Aïcha. 1988. *La Salaire de Madame.* Paris: Editions Hazan.

Benali, Driss et al. Forthcoming. *Pour une stratégie nationale de développement social integré.* UNDP Report.

Ben-Meir, Jason. 2004. "Create a new era of Islamic-Western relations by supporting community development," *International Journal on World Peace,* Vol. 21, No. 1, March 2004, pp. 43–60.

Benhaddou, Ali. 1997. *Maroc: Les Elites du Royaume.* Paris: L'Harmattan.

Benanni-Chraïbi, Mounia. 1994. *Soumis et rebelles: les jeunes au maroc*. Casablanca: Editions Le Fennec.

Bhagwati, Jagdish. 2005. *In Defense of Globalization*. Oxford: Oxford University Press.

Bougroum, Mohammed and Ibourk, Aomar. 2003. "The effects of job-creation schemes in Morocco," *International Labour Review*, Vol. 142, No. 3, pp.341–371.

Brothers, Robyn. 2000. "The computer-mediated public sphere and the cosmopolitan ideal," *Ethics and Information Technology*, Vol. 2, No. 2, June 2000, pp. 91–97.

Cardoso, Frederique and Faletto, Enzo. 1979. *Dependency and Development in Latin America*. Berkeley, CA: University of California Press.

Cohen, Shana. 2004. *Searching for a Different Future*. Durham, NC: Duke University Press.

Cohen, Shana. 2003. "Alienation and Globalization in Morocco," *Comparative Studies in Society and History*. Vol. 45, No. 1, pp. 168–169.

Crawford, David. 2002. "Morocco's invisible Imazighen," *Journal of North African Studies*, Vol. 7, No. 1. Spring 2002. pp. 53–70.

Crewe, Emma and Harrison, Elizabeth. 1999. *Whose Development?: An Ethnography of Aid*. London: Zed Books.

Dalle, Ignace. 2001. *Maroc 1961-1999: L'espérance brisée*. Paris: Maisonneuve & Larose.

Daoud, Zakya. 1997. *Marocains des Deux Rives*. Paris: Editions Ouvrières.

Dubet, François. 2004. "Between a defence of the subject and a politics of the subject," *Current Sociology*, Trans. Amy Jacobs, Vol. 52, No. 4, (July) pp. 693–716.

El Aoufi, Noureddine. 1990. *La Marocanisation*. Casablanca: Editions Toubkal.

El Ouadie, Salah. 2001. *Le marié*. Casablanca: Editions Tarik.

Escobar, Arturo. 1994. *Encountering Development*. Princeton, NJ: Princeton University Press.

Evans, Peter. 1979. *Dependent Development*. Princeton, NJ: Princeton University Press.

Evans, Peter. 2001. *Livable Cities? Urban Struggles for Livelihood and Sustainability*. Berkeley, CA: UC Press.

--------. 1996. *Embedded Autonomy*. Princeton, NJ: Princeton University Press.

Ferguson, James. 1994. *The Anti-Politics Machine*. Minneapolis, MN: University of Minnesota Press.

Ferrari, Laurent with Tardieu, Bruno. 2000. "No More Power Cuts: Electricity is a Public Service," in *Artisans of Democracy*, Lanham, MD: University Press of America.

Freeman, Amy, 2004. "Re-locating Moroccan women's identities in a transnational world: The 'woman question' in question," *Gender, Place, and Culture*, Vol. 11, No. 1, March 2004, pp. 17–41.

Granovetter, Mark and Swedberg, Richard. 2001. *The Sociology of Economic Life*. Boulder, CO: Westview Press.

Huntington, Samuel. 1968. *Political Order in Changing Societies*. New Haven: Yale University Press.

Inkeles, Alex and Smith, David. 1974. *Becoming Modern: Individual Change in Six Developing Countries*. Cambridge, MA: Harvard University Press.

Inkeles, Alex. 1983. *Exploring Individual Modernity*. NYC. Columbia University Press.

Ketterer, James. 2001. "Networks of discontent in northern Morocco: Drugs, opposition, and unrest," *MERIP*, pp. 30-33, 45.

Laâbi, Abdellatif. 2001. *Les Rêves sont têtus: Ecrits politiques*. Paris: Eddif.

Lahbabi, Mohammed. 1970. *Les Années 80 de Notre Jeunesse*. Casablanca: Editions Maghrebines.

Laroui, Abdallah. 1977. The History of the Maghrib: an interpetive essay. Princeton, NJ. Princeton University Press.

Laroui, Abdellah. 1993. *Esquisses Historiques*. Casablanca: Centre Culturel Arabe.

Les Indicateurs Sociaux. 1993. Rabat: Direction de la Statistique.

Maddy-Weitzman, Bruce. 2001. "Contested identities: Berbers, 'Berberism' and the state in North Africa," *Journal of North African Studies*, Vol. 6, No. 3. pp. 23–71.

McKibben, Bill. 2005. "Poor no more," *Christian Century*, May 31, 2005.

Mander, Jerry and Goldsmith, Edward. 1996. *The Case Against the Global Economy*. San Francisco: Sierra Club Books.

Marzouki, Ahmed. 2000. *Tazmamart Cellule 10*. Casablanca: Editions Tarik.

Mazur, Robert E. 2004. "Realization or deprivation of the right to development under globalization?: Debt, structural adjustment, and poverty reduction programs." *GeoJournal*, Vol. 60, No. 1.

Mernissi, Fatima. 1997. *Les Aït Débrouille Haut-Atlas*. Casablanca: Editions Le Fennec.

Molavi, Afshin. "Supporting the Private Sector," in "The Knowledge Economy," *Development Outreach*, Washington, D.C.: World Bank, Fall 2001.

Msefer, Assia Akesbi. 1998. *Ecoles, Sujets, et Citoyens*. Casablanca: Edit Consulting.

Munson, Henry. 1993. *Religion and Power in Morocco*. New Haven, CT: Yale University Press.

Narayan, Deepa et al. 2000. *Voices of the Poor: Can Anyone Hear Us?*. Oxford: Oxford University Press.

Nussbaum, Martha. 1999. "Women and equality: The capabilities approach," *International Labour Review*, Vol. 138, No. 3.

Offenheiser, Raymond C. and Halcombe, Susan H. 2003. "Challenges and opportunities in implementing a rights-based approach to development: An Oxfam America perspective," *Nonprofit and Voluntary Sector Quarterly*, Vol. 32, No. 2, pp. 268-301.

Ossman, Susan. 2002. *Three Faces of Beauty*. Durham, NC: Duke University Press.

Rodrik, Dani. 2006. "Goodbye Washington consensus, hello Washington confusion," *Journal of Economic Literature*.

Roque, Maria-Angels, ed. 2004. *La Société Civile au Maroc*. Paris: Editions Publisud.

Rosenfeld, Jona M. and Tardieu, Bruno. 2000. *Artisans of Democracy*. New York: University Press of America.

Samuels, David. 2004. "From socialism to social democracy," *Comparative Political Studies*, Vol. 20, No. 10, pp. 1-26.

Sen, Amartya. 1999. *Development as Freedom*. New York: Knopf.

Stiglitz, Joseph. 2002. *Globalization and Its Discontents*. New York: W.W. Norton.

Stubbs, Paul. 2003. "Znknational non-state state actors and social development policy," *Global Social Policy*, Vol. 3, No. 3, pp. 319–348.

Sweet, Cathy. 2001. "Democratization without democracy: Political openings and closures in modern Morocco," *Middle East Report* 218, Spring 2001.

Touraine, Alian. 2004. "On the frontier of social movements," *Current Sociology*. Vol. 52, No. 4 (July). pp. 717–725.

UNDP. 2002. *Arab Human Development Report 2002*.

UNICEF. 2004. *State of the World's Children*.

Van de Walle, Dominique. 2004. "Do Basic Services and Poverty Programs Reach Morocco's Poor?: Evidence from Poverty and Spending Maps," World Bank Discussion Paper, November 2004.

Vermeren, Pierre. 2002. *Ecole, elite et pouvoir au Maroc et en Tunisie au XX siècle*. Rabat: Alizés.

World Bank. 2003. *Better Governance for the Middle East and North Africa*.

World Bank. 2005. *Project Information Document (PID) Appraisal Stage*. Report No.: AB1111.

Yassine, Abdessalam. 1998. *Islamiser la modernité*. Al Afok Impressions.

Yassine, Abdessalam. 2000. *Winning the Modern World for Islam*. Iowa City, IA: Justice and Spirituality Publishing.

Zeghal, Malika. 2005. *Les Islamistes Marocaines: De Défi á la Monarchie*. Paris: La Découverte.

Index

A

Activism, 61, 112, 155, xviii
    social, 121–125
AIDS, 25, 106
Al-adl wa al-ihsan, 52, 53, 54, 102, 110
    agility and elusiveness, 111
    mobilization of students, 108
    popularity of, 85
    position of women and, 88
al-Amana, 133
Amazigh
    activists, 4, 93, 95, 102
    autonomy, 92–96
    festival, 94
    language and culture, 86, 93
        in education, 94
    marginalization, 102
    movement, 96
    organizations, 92, 95
    political party, 95
    population in Morocco, 92
    in rural areas, 93
AMSED, 25, 28

Association Marocaine des Droits
    Humains, 25
Association Migrations
    Développement, 123
ATTAC, 19, 24
Automobile sales, 136
Awdellil, 73–75

B

Basic Education Reform Support
    Program, 130
Beijung Conference on Women
    (2000), 88
Berber. See Amazigh
Birth rate, 41, 119
Bouznika (Morocco) Social
    Reform, 19

C

Catholic Relief Services, 25
Children, 92, 108
    custody of, 91
    education of, 84, 129, 143, 150
    health, 25, 106

mortality, 27, 41
NGOs, 24
in outlying areas, 129
Civil society organizations, 85
Collectif Printemps de l'Egalité, 90
Comprehensive Development
    Framework, 130
Conseil Consultatif de Droits de
    l'Homme, 61
Conseil de la Jeunesse, 60
Conseil Economique et Social, 61
Conseil Supérieur, 61
Conseils Régionaux des Ouléma,
    58, 61
Country Assistance Strategy for Morocco,
    115, 119
Cyberpirates, 142, 148

D

Derb Ghallef, 138
Diploméschômeurs, 82
Divorce, 89, 91
Domestic abuse, 105
Drug cartels, 34

E

Education, 119, 127, 138, 145
    AIDS, 25, 106
    Amazigh languages in, 94, 95
    Arabic, 93
    arabicization in, 56
    basic, 129
    constitutional right to, 100
    employment and, 104, 129
    expansion of, 31, 59
    female, 25, 106, 120, 130

male vs., 27
foreign vs. Moroccan, 148
French, 93
higher, 144, 145
improving quality of, 70
indicators, 120
investment in, 10, 129, 140
Islamic, 103
literacy and, 120
parallel, 108
participation in global market
    and, 137
political institutions and, 52
primary, 119, 167
private, 86, 143, 150
    public vs., 150
privatization of, 86
progress in, 35, 37
public, 141, 143, 144
    expenditure on, 160, 167
    private vs., 150
quality of, 118
rates, 118
reform, 130
religious, 72, 108
rural, 127
status of, 117
tertiary, 167
unemployment and, 139
workshops, 163
World Bank investment in, 129
Employment, 30, 45, 77, 82, 130,
    142, 153, 155. See also
    Unemployment
as developmental objective, 119
education and, 104, 129
opportunities, 17

rates, 115
temporary, 69, 144
white-collar, 126
Entrepreneurs, 139, 148

F

Family code. See Moudawana
Female literacy. See Women,
        literacy

G

Globalization
    benefits, 148
        economic, 17, 154
    critique of, 104
    cultural, 86, 149
    cultural consumption and, 135
    development policy, 35–45
    as development strategy, 113,
        115
    economic, 17, 148, 154
    entrepreneurship and, 139
    establishing existential space
        within, 134
    identity and, 147
    of liberal democracy, 66
    multiple voices, 15–30
    poverty and, 19
    quality of life and, 52, 155
    role of state in, 30–35
    social structure and, 17–18, 52,
        115, 155
    technological, 55
Grameen Bank, 25, 28

H

Health, 130, 145, 153
    consulting firms, 24
    education, 106
    environment and, 24
    facilities, 131, 146
    government expenditure on,
        146, 160n
    indicators, 117
    market reform and institutions
        for improvement of, 72
    maternal, 43, 117
    Ministry of, 43
    programs, 19, 25, 43
    social equality and, 147
HIV/AIDS, 25, 106
Hospitals, 146
Human development index, 120

I

IMF. See International Monetary
        Fund (IMF)
Institut Royal de la Culture
        Amazighe, 94
Institutional reform, 121–125
Intellectuals, 81
International Finance Corporation,
        136
International Monetary Fund
        (IMF), 24, 29, 30, 121,
        136
    advocacy for human rights, 12
    analyses, 115
    criticism, 142
    evaluations, 116
    market strategies, 153

Morocco as success story for, 116
recommendations, 34, 37
World Bank *vs.*, 16
Islam, 93
    conservative, 83, xviii
    democracy and, 103
    education and, 103
    modernity *vs.*, 96, 103, 104
    perspectives on women, 108
    politics and, 54, 99, 109–111
    radicalized, 109–111
    secularism *vs.*, 95
    social cohesion and, 107–108
    Sufi, 58, 108
Istiqlal, 59

J

Jeunes Entrepreneurs/Promoteurs, 122
John Snow, Inc., 24
Juridique et Psychologique pour Femmes Agressées, 106
Justice and Development Party, 88, 91

K

King Mohammed VI, 2, 5, 50, 55, 127
    advocacy for women's rights, 89
    creation of Institut Royal de la Culture Amazighe, 94
    interview with, 157n
    political liberalization and, 59–62
    role of, 1, 2, 112

L

Le Journal, 66, 163n
Le Journal-Hebdomaire, 161n, 162n
Literacy, 69
    classes, 105
    female, 25, 43, 91, 105, 117, 120, 124, 127, 129
    legal, 163
    major campaign, 119
    program, 26, 43
    quality of life and, 129
    workshops, 90

M

Maraboutisme, 58
Market liberalization, 138, 139
Marriage, 70, 105
Media, 8
    Arab, 6
    global, 111
    independent bodies, 43
    interviews, 30
    liberal, 8
    Moroccan, 7, 8
    political use of, 89–90
    range of, 42
    technology, political effect of, 111–112
    Western, 5
    of youth culture, 8, 73–75
Medical care, 146
Middle class, 36, 53, 59, 137
    consumer orientation, 136
    establishment of, 31
    global, 148
    growth of, 136

identification of, 150
identity, 135
intellectuals, 81
managers of development, 150
occupations, 141
support for, 148
transformation, xii
Millennium Development Goals,
        17, 26, 27
Mohammed VI. See King
        Mohammed VI
Morocco
    birth rate, 41
    cultural-political movements
        public sphere and, 96–109
    education, xi
    historical perspectives, 1–10
    industrialization, xv
    issues of contestation, 86–96
        Amazigh autonomy, 92–96
        family, 86–91
    literacy, xi
    market reform and, 37–41
    modernization, 35–37
        to universalism, 10–15
    from modernization to
        universalism, 10–15
    political liberalization, 41–44
        process of, 55–62
    postcolonial, political
        radicalization and
        mobilization in, 82–86
    primary school completion
        rates, xi
    World Bank assessment of, 32,
        39, 142
Mortality, 27, 43, 119

Moudawana, 70
    as document of female
        bondage, 87
    drafting of, 87
    old vs. new, 87, 88
    reform, 51, 86, 87, 90, 91
Music, 7, 134, 135
    as form of resistance, 135
    global recording industry, 134
    mixing, 135
    suppression of, 9

N

Nation-state identity, 135
National Initiative for Human
        Development, 39, 131,
        152
National School of Engineering,
        145
NEPC, 20, 21
Non-governmental organization
        (NGO), 123
    collapse of projects, 121
    social services, 28
    state partnership with, 25

O

Oxfam, 12, 14

P

Politics, 52, 55–62, 161n
    among youth, 71, 72
    conventional, 71, 72, 76
    cultural debates and, 81,
        96–109

drug cartels and, 34
global, 81, 88
identity and, 71
interest in, 67
media technology and, 111–112
militants and, 95
mobilization of masses and, 85
movements associated with,
    79–112
negative attitudes toward, 72
participation in, 67, 81, 153
role of monarch in, 2, 112
transnational pan-ethic, 96
urban, 52
women and, 68, 85
Polygamy, 91
Poverty, 115, 118
borderline, 44
downward trends, 119
globalization and, 19
NGOs and alleviation of, 113
participatory assessments, 113
persistence of, 44
rate, 39, 44, 116
rural, 119
unemployment and, 44, 115
urban, 116
    rural *vs.*, 39
Poverty Eradication in Rural Areas,
    130
Prisoners, 84
Professions, 141

Q

Quality of life, 52, 115–121, 155

S

Salafiyya Jihâdiyya, 109, 111
Save the Children, 24, 25, 31
Shanghai Conference (2004), 15
Silicon underground market, 138
Social activism, 121–125
Social identity, 133–149
Social space, transformation of,
    125–133
Social stratification, 136–147
Sufi Islam, 58, 108

T

Teachers, 141. See also Education
Tel Quel, 8, 66, 73, 157n, 158n,
    161n, 162n, 164n
Tobin Tax, 19
Transparency International, 85

U

Unemployment, 16, 45, 69, 138
among educated youth, 39, 122,
    139
corruption and, 31
decrease, 45, 69
economic growth and, 118
economic liberalization and, xii
education and, 39, 139
indifference to, 67
job preparation and, 144
long-term, 139
massive, 19
poverty and, 44, 115
rates, 39, 100, 139
reducing, 45, 116
rise in, 138

social space and, 135
university as factory of, 144
urban, 39, 116, 118, 139
World Bank approach to, 118
UNICEF, 129
Union Socialiste de Forces
        Populaires (USFP), 42,
        51, 70, xi
    leadership, 65
    opposition, 59
United Nations Development
        Program (UNDP), 133
United States Agency for
        International
        Development (USAID),
        22

V

Voting age, 68

W

Women
    appointment to political
        positions, 89
    as custodial parents, 91
    domestic abuse, 105
    employment, 118, 142
    empowerment, 26
    as entrepreneurs, 122
    Islamic perspectives on, 108
    legal resources for, 106
    literacy, 25, 43, 91, 105, 117,
        120, 124, 127, 129
    medical care, 146
    mortality rates, 43
    Moudawana and, 87

organizations, 90, 92, 105
politics and, 68, 85
programs for development, 90
rights, 87–91
in rural communities, 128
seeking public sector jobs, 100
solidarity among, 106
status of, 3, 34, 43
in support of political
        prisoners, 84
World Amazigh Congress, 92
World Bank, 6, 15, 24, 28, 90, 114
    advocacy for human rights, 12
    approach to unemployment,
        118
    assessment of Morocco, 32, 39
    Comprehensive Development
        Framework, 130
    critics, 22, 34
    e-discussion, 19
    evaluations, 116
    IMF vs., 16
    investment in education, 129
    outlook for Morocco, 40
    recommendations, 9, 104
    reports, 115, 116, 117
    research teams, 121
    strategy to reduce
        unemployment, 118
    world vision, 18
World Social Forum, 24
World Vision, 24